NOBEL PRIZE CONVERSATIONS

Also published in the
Saybrook "Human Future" Series:

My Quest for Beauty, by
Rollo May

*American Politics and
Humanistic Psychology,*
by Rollo May,
Carl Rogers and Others

Fully Alive, by
Roy Laurens

NOBEL PRIZE CONVERSATIONS

with Sir John Eccles
Roger Sperry
Ilya Prigogine
Brian Josephson

WITH A COMMENTARY BY
NORMAN COUSINS

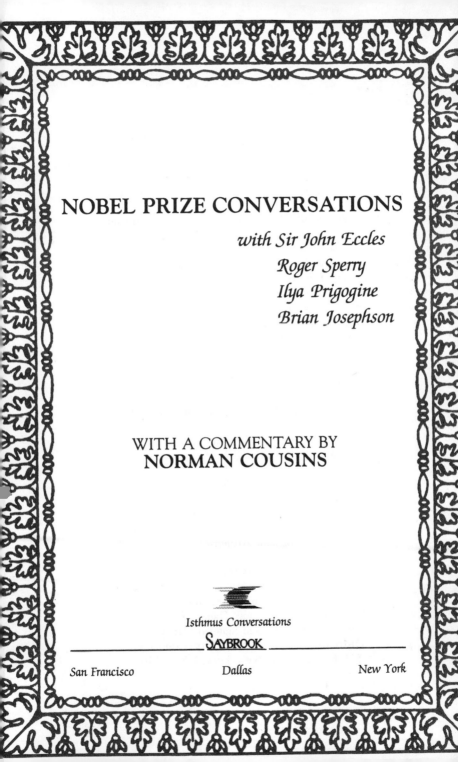

Isthmus Conversations

SAYBROOK

San Francisco   Dallas   New York

Designed by Alison Esposito and Chiles & Chiles

Library of Congress Cataloguing in Publication Data

Nobel Prize conversations with Sir John Eccles, Roger Sperry, Ilya Prigogine, Brian Josephson.

(Isthmus conversations)
1. Scientists—Interviews. 2. Science—Philosophy.
3. Science—Social aspects. 4. Nobel prizes. I. Eccles, John C. (John Carew), Sir, 1903–   . II. Cousins, Norman. III. Series.
Q141.N68 1985 501 85-61685
ISBN 0-933071-02-7

Saybrook Publishing Company
3518 Armstrong Avenue, Dallas, TX 75205
Printed in the United States of America
Distributed by the W.W. Norton Company
500 Fifth Avenue, New York, NY 10110

DEDICATION

The Isthmus Institute exists because of the dreams of the leading citizens of Dallas, Texas for a better future for all humankind. Among the men who were particularly respon-sible for bringing the Nobel Laureates to Dallas are Dr. James Hall, president of the Isthmus Institute at the time; Dr. Albert Outler, a great scholar known for his definitive volumes on the life of John Wesley; and C.A. Rundell. The Nobelstiftelsen of Stockholm deserves special recognition for its gracious cooperation in the preparation of this book and for its outstanding record of service to the highest interests of humankind.

DEDICATION

A WORD OF THANKS

One of the most prominent editors of our times, Norman Cousins, showed us how to tune this book and tighten it up. Dr. Hugo Rossi graciously consented to read the manuscript, and his assistance was invaluable. Nathan Mitchell, senior editor of Saybrook Publishing, is responsible for the prelude, introductions, commentaries and the first interlude. Without him the work would not have been possible. Frank O'Connor was production editor, Carolyn Smith prepared the manuscript, and Joan Howell helped everyone work together.

CONTENTS

PRELUDE

The Karolinska Institutet
The Isthmus Institute   3

CONVERSATION ONE

Mind Enfolds Brain:
The Mentalist Revolution   37

CONVERSATION TWO

Mind Enfolds Matter:
Evolution and God   71

CONTENTS

INTERLUDE I
To Think of Time   103

INTERLUDE II
The Rediscovery of Time   117

CONVERSATION THREE
The Future of Humans
Is the Humans of the Future   149

POSTLUDE
Commentary by
Norman Cousins   185

NOTES   197

ACKNOWLEDGMENTS   207

NOBEL PRIZE CONVERSATIONS

I know if I find you I will have to leave the earth
and go on out
    over the sea marshes and the brant in bays
and over the hills of tall hickory
and over the crater lakes and canyons
and on up through the spheres of diminishing air
past the blackset noctilucent clouds
    where one wants to stop and look
way past all the light diffusions and bombardments . . .

And I know if I find you I will have to stay with the earth
inspecting with thin tools and ground eyes
trusting the microvilli sporangia and simplest
    coelenterates
and praying for a nerve cell
with all the soul of my chemical reactions . . .

From "Hymn,"
by A. R. Ammons

PRELUDE

ತನ

THE NOBELSTIFTELSEN
THE ISTHMUS INSTITUTE

ತನ

FROM STOCKHOLM TO DALLAS

INTRODUCTION

The Nobel Prize Conversations which form the nucleus of this book are introduced with lines from the poem "Hymn," by contemporary American poet A. R. Ammons. The poet invites us to reflect on the double character of human nature, earth-bound yet propelled toward the farthest reaches of the universe. Humans, by definition, must simultaneously "leave the earth" to climb "past all the light diffusions," yet "stay with the earth . . . trusting the microvilli sporangia."

The reflections prompted by "Hymn" beckon us to

consider specific times and places, where we will meet the four men whose research was rewarded at the annual Nobel Prize ceremonies in Stockholm: Sir John Eccles (1963, Physiology or Medicine); Dr. Brian Josephson (1973, Physics); Dr. Ilya Prigogine (1977, Chemistry); and Dr. Roger Sperry (1981, Physiology or Medicine). From their earth-bound laboratories in various parts of the world—England, Australia, Belgium, Texas, California—these scientists have followed A. R. Ammons's advice by reaching out with their minds "past the blackest noctilucent clouds," while faithfully "inspecting with thin tools and ground eyes" the simplest coelenterate, the tiniest nerve cell.

It was not chance which brought these four men to the Isthmus Institute meetings in Dallas, Texas, in November of 1982. Each of them possessed one piece of a mosaic which, put together, might speak to the future of humanity. They came from great distances and at some cost to themselves in the interest of their fellow human beings. Although each represented a different scientific discipline, they had one thing in common: each had received the Nobel Prize, each had used the gifts of intelligence in the service of human life. The awards they had received in ceremonies at Stockholm's Nobelstiftelsen signified that the scientific community as a whole had come to accept their work not only as valid but as important to humankind.

Another element also unites these four Nobel Laureates. Each of them is concerned about the relation between human mind and human brain, about the role of human consciousness in an evolving universe, about the interplay between time and mind, about the world as a "work of art" which cannot simply be reduced to neural events within the brain or to immutable mechanisms measured by quantum analysis. Like the devoted inspectors suggested by the images in Ammons's "Hymn," these brilliant researchers have left no stone unturned in their search for human tracks in a universe that is often viewed as a mindless automaton rushing nowhere at unbelievable speed.

As Norman Cousins writes in his concluding commentary on the conversations collected in these pages, "This book is a superb effort in convergence and coordination. It deals with the furthest reaches of modern scientific theory, but does so by connecting new findings and theories to one another and by creating a pattern rather than a splattering of colors." Cousins notes, again, that the thinking represented here exemplifies "the reaching-out of highly differentiated intellects to one another."

Cousins's comments are important as the reader begins to engage in his or her own "conversation" with the Nobel Laureates whose views are brought together here. In science, as in literature and the arts, convergence does not mean unanimity, nor does coordination signify agreement. Even when scientists reach similar conclusions in the laboratory, they may still adhere to different philosophies, worldviews and versions of human nature. Readers are encouraged to look for dissimilarities as well as points of coordination in the conversations collected here. As Cousins reminds us, these scientists have "highly differentiated intellects" and their reaching out to one another creates a "pattern" composed of many *different* threads.

All these Laureates are vitally concerned about the

human future and about the role of human mind in that future. But while their concern is common, their individual views differ, sometimes markedly. Though a pattern of convergence may be emerging, we are still far from a unified concept of the human future to which these four scientists would subscribe.

For many of us the Nobel Prize, whether awarded in the arts, the sciences, or humanitarian causes, is something of an abstraction. Partly because few of us have ever seen the ceremonies or met a Nobel Prize recipient, the events and the winners seem remote. It is easy to forget that the persons honored by the Nobel Foundation are real flesh-and-blood, that the chosen honorees have often been controversial in private, public and political life, and that every Nobel Prize winner joins company with a host of predecessors with whom he or she may not feel entirely comfortable.

The Nobel Laureates represented in this book call forth a response not only to their ideas and intellectual achievements, but to their personal qualities as well. Eccles, Josephson, Prigogine and Sperry are more than dispassionate observers of nature; they are passionately involved in its mysteries, and compassionately concerned about the relation between the conclusions of science and the quality of life. Science, for them, is not sheer detached measurement or observation; it invites—and demands—attention to ethics, to the condition of our biosphere, to systems of belief both religious and secular, and to the impact of research on our future.

Before launching into the conversations conducted in 1982 at the Isthmus Institute in Dallas, readers are invited

in this "Prelude" to meet the Nobel personalities themselves, to get a feel for their human reality, and to sense
some of the excitement and wonder that surround the Prize
ceremonies in Stockholm. Though these four men were not
all honored in the same ceremony, a bit of imaginative
liberty may be taken to suggest the flavor of what happens
each year at the Nobelstiftelsen.

وة‍‍   ‍‍‍هم

The locale is Stockholm: the legendary beauty of its waterways, its statuesque blondes, its charm, its midnight suns,
its Volvos, its high standard of living, its higher suicide rate,
its enlightened socialism, Bergman films. And the old hall
where the Nobel Prizes are given, the world's nerve center
for those who seek to know so that they can serve. The dark
walls brood, suspended between nightmare and sanctity, as
in the silent chessgame between Death and the disillusioned
old knight in Bergman's "Seventh Seal." The holy family
of three travellers: Karin reading the Apocalypse at table,
the cracking knock of Death at the door, the black beating
wings of the last angel.

Bergman's angels lack the assurance of those presences
that hide just behind the scrim of conciousness in Rilke's
Elegies. Bergman's are icier, stranger, unsure whether or
not they are fallen—and thus infinitely more dangerous. A
habitation of angels, irridescent presences: that is the hall of
the Nobelstiftelsen, with its soft swirling light.

One is led to contemplate the great men and women who

have come here since 1901 to be honored. Hemingway floats by, his face blotched from alcohol, his eyes bright but fugitive. And naturally Einstein is there, a slight man, stooped, his head barely able to support its great wings of hair. "Subtle is the Lord," he mutters, "subtle, but not malicious." And swifter than light, he disappears into his preoccupations.

Other Nobel Prize winners enter into one's contemplation of this ancient hall. There is Nelly Sachs, poet of the Holocaust, chronicler of this century's darkest time. She can't weigh a hundred pounds, and even at 70, her skin has the pearly lustre of light escaping slowly from Stockholm's sky in summer. As always, she chants her sad choruses:

> O the chimneys
> On the ingeniously devised habitations of death
> When Israel's body drifted as smoke
> Through the air . . .

Her voice has the quality of an oboe, bewitching, penetrating. She hovers above Hemingway a moment, as though she senses in him a loss akin to the pain she feels. Then, Mother Theresa: she seems ill at ease as she sits nervously pleating the blue border of her white cotton sari. How like her. She *would* rather be sitting on a stoop in some infested alleyway, holding a dying child. Theresa, Nelly Sachs: Rachael weeping for her children because they are no more.

The light in the great hall intensifies and banishes contemplation. The Swedish king is now entering the assembly.

The sound of the careful, slow Nordic sifting of men mixes with the scent of the hall's rich wood panelings. The scene becomes charged, transfigured. The king, actually a kind, intelligent man, looms grim as a Norse god. We are in Valhalla.

Let us pause in our meditation to describe those particular Nobel Prize winners who, after being honored in Stockholm, came to the Isthmus Institute in Dallas. There is Roger Sperry, who looks so much like a saint. It is the combination of his white hair and beard with his calm, unlined face. Perhaps it is also his age. He is 71. Coming to the Isthmus Institute took a deep commitment to the future of his fellow humans and also took considerable personal courage. He suffers from a slow form of the same progressive disease which afflicted Lou Gehrig. However, he has chosen to live all the way and contribute to the limit of his physical ability. His intellectual clarity and passion for human values shine through his words. Sometimes speaking becomes difficult for him and some of his remarks recorded here are actually spoken by his wife.

Brian Josephson is young, lean, active, with curly brown hair. He received the Prize in physics at the age of 33. Something slightly unruly and mischievous plays on Josephson's face, in contrast to Sir John Eccles, who epitomizes the kind of elegance associated with the British aristocracy—quiet urbanity, serene self-control. There's no doubt that Sir John, whose home is now Switzerland, lives in a way that comports well with his elegant style of dress and speech.

Compare these three men and observe how distinctive each of them is. Here is Josephson. Think of what it must be like to hold the world's most prestigious honor at the age of 33. Sperry, on the other hand, reflects the sun and sea of southern California. In true Californian style he combines a radical openness to the experimental and new with a profound reverence for the life-giving land. In Eccles, centuries of wisdom linked to British culture and stretching back to the time of Caedmon and Bede the Venerable, are brought into focus.

Josephson looks bemused, like a junior Cambridge don who must learn to suffer undergraduates gladly. In him is the dual atmosphere of Cambridge—high-jinks and hard studies, while the River Cam flows endlessly between antique towers. Recognition has come to him so early. With Sperry it was different. For thirty years he worked quietly on one clearly focused question, "What is human mind?" His answer is clear and simple. "Mind controls matter, is superior to brain in its capacity to will, intend, command and direct." Eccles, too, a scholar of senior age like Sperry, has invested many years of research to show how the non-material realities of mind and mental intention act upon the material brain. Both Sperry and Eccles have labored long to overcome the prejudices of a materialist science which has sought to reduce human mind to the brain's electrical system and chemistry.

But our imaginations are pulled back once more to the Nobel Prize ceremonies themselves. The Nobel Hall, though darkly panelled, is illuminated by bright planes of light throwing sharp contrasts. The dust motes dancing in the light might well be the same ones present when Einstein or Niels Bohr or Max Planck received their awards. Maybe this place truly is inhabited by a dust cloud of personalities. Ernest and Nelly. Alfred, Niels and Max. Saul of Chicago. Sigrid of Oslo. Somehow they are all really here, meeting, crowding the roofbeams.

Let us listen to the award speeches which introduced the work of Nobel Laureates Eccles, Josephson, Prigogine and Sperry. Professor David Ottoson speaks first, to present the Prize to Roger Sperry:

(Photograph of Roger Sperry, overleaf)

Roger Sperry

PROFESSOR OTTOSON

Your Majesties, Your Royal Highnesses, Ladies and Gentlemen,

One day in October, 1649, René Descartes, the French philosopher and mathematician acknowledged as the greatest brain researcher of the period, arrived in Stockholm at the pressing invitation of Queen Christina. It was with much hesitation that Descartes came to Sweden, "the land of bears between rocks and ice." In the letters to his friends, he complained bitterly that he was obliged to present himself at the Royal Palace at five o'clock each morning to instruct the young queen in philosophy, so avid was she for knowledge. Modern brain research scientists and followers in the Cartesian footsteps are not faced with the same demands as winners of the Nobel Prize, but they are met with other tribulations, expectations.

Descartes with the help of philosophy sought to find the answer to his questions of the functions of the mind. Later research has had other means at its disposal and has tried to feel its way forward by other methods. Sperry has succeeded with sophisticated methods in extracting from the brain some of its best-guarded secrets and has allowed us to look into a world which until now has been nearly completely closed to us.

The brain consists of two halves, hemispheres, which are structurally identical. Does this mean that we have two brains or that the two hemispheres have different tasks? The answer to this question can appear impossible to find because the brain halves are united by millions of nerve threads and therefore work in a complete functional harmony.

However, it has been known for more than a hundred years that despite their similarity and close linkage the two hemispheres have in part different tasks to fulfill. The left hemisphere is specialized for speech and has therefore been considered absolutely superior to the right hemisphere. For the right hemisphere it has been difficult to find a role and it has generally been regarded as a "sleeping partner" of its left companion. In a way the roles of the two hemispheres were somewhat like those of man and wife of an old-time marriage.

In the beginning of the 1960s Sperry had occasion to study some patients in whom the connections between the two hemispheres had been severed. The surgical intervention had been undertaken as a last resort to alleviate the epileptic seizures from which the patients suffered. In most of them an improvement occurred and there was a decrease in the frequency of their epileptic fits. Otherwise, the operation did not appear to be accompanied by any changes in the personality of the patients. However, Sperry was able, using brilliantly designed test methods, to demonstrate that the two hemispheres in these patients had each its own stream of conscious awareness, perceptions, thoughts, ideas and memories, all of which were cut off from the corresponding experiences in the opposite hemisphere.

The left brain half is, as Sperry was able to show, superior to the right in abstract thinking, interpretation of symbolic relationships and in carrying out detailed analysis. It can speak, write, carry out mathematical calculations and in its general function is rather reminiscent of a computer. Furthermore, it is the leading hemisphere in the control of the motor system, the executive and in some respects the aggressive brain half. It is with this brain half that we communicate. The right cerebral hemisphere on the other hand is mute and in essence lacks the possibility to reach the outside world. It cannot write and can only read and understand the meaning of simple words in noun form and does not grasp the meaning of adjective or verb. It almost entirely lacks the ability to count and can only carry out simple additions up to twenty. It completely lacks the ability to subtract, multiply and divide. Because of its muteness, the right brain half gives the impression of being inferior to the left. However, Sperry in his investigations was able to reveal that the right hemisphere in many ways is clearly superior to the left. Foremost, this concerns the capacity for concrete thinking, the apprehension and processing of spatial patterns, relations and transformations. It is superior to the left hemisphere in the perception of complex sounds and in the appreciation of music; it recognizes melodies more readily and also can accurately distinguish voices and tones. It is, too, absolutely superior to the left hemisphere in perception of nondescript patterns. It is with the right hemisphere we recognize the face of an acquaintance, the topography of a town or landscape earlier seen.

It is soon fifty years since Pavlov, the great Russian physiologist, put forth the suggestion that mankind can be divided into thinkers and artists. Pavlov was perhaps not entirely wrong in making this proposal. Today we know from Sperry's work that the left hemisphere is cool and logical in its thinking, while the right hemisphere is the imaginative, artistically creative half of the brain. Perhaps it is so that in thinkers the left hemisphere is dominant whereas in artists it is the right.

(Professor Ottoson turns to speak to Roger Sperry directly:)

Dr. Sperry, you have with your discoveries written one of the most fascinating chapters in the history of brain research. You, Dr. Sperry, have with your studies given us more profound insights into the higher functions of the brain than all the knowledge acquired in the eighteenth century.

It is a privilege and pleasure for me to convey to you the warmest congratulations of the Nobel Assembly of Karolinska Institutet and to invite you to receive your Nobel Prize from the hands of His Majesty the King.

COMMENTARY

As Professor Ottoson concludes his remarks, one's attention shifts back to Sperry. His eyes, so alert, are intensely trained on all that's happening. You can tell he relishes this moment. Mother Theresa would recognize a kindred spirit. Sperry is a rather small man who is making his way across the stage toward King Gustav, who kindly takes extra steps to meet him more than half way. Watching this scene, thinking of Sperry's conviction that the mind is *more* than brain, seeing the light glisten on his snowy white beard, one recalls lines from Wallace Steven's poem "The Snow Man:"

> One must have a mind of winter
> To regard the frost and the boughs
> Of the pine-trees crusted with snow . . .

. . . the listener, who listens in the snow,
And, nothing himself, beholds
Nothing that is not there and the nothing that is.

Of this poem Stevens himself wrote that he wished it to be
an example of the necessity of identifying oneself with reality in order to understand and enjoy it. Sperry's work as a
scientist has been similar: to identify with the reality of
human mind, its outreach, ideas and intentions, in order to
appreciate more fully its causal control over the brain's
neuronal traffic. Perhaps that is what Professor Kjell Fluxe
of Stockholm's Karolinska Institutet meant when he remarked that Roger Sperry's work shows how man has a
soul.

Sperry and his wife leave their brief meeting with the
king slowly, smiling. Aside from all the scientific brilliance
that he has just demonstrated, Sperry's face radiates love.
Here are two people who simply and obviously care for
each other. Life has lost none of its attraction for them. *Iam
virescunt.*

Sperry makes an unforgettable impression. His very presence seems to enflesh all that his research has discovered.
More poetry comes to mind, lines from William Butler
Yeats's "Sailing to Byzantium." For Yeats, himself a winner
of the Nobel Prize for literature in 1923, Byzantium stood
as symbol for the triumph of spirituality as expressed in art.
The poem plays on the paradoxical situation of youth and
old age: how ancient is the young's desire for love and
passion; how youthful is an old man's spirit soaring to the
heights of a wisdom only age can teach. Yeats wrote:

> An aged man is but a paltry thing,
> A tattered coat upon a stick, unless
> Soul clap its hands and sing . . .

Sperry's soul, clapping its hands, singing, as he walks hand-in-hand with his wife.

                       ❧   ☙

In the old hall of the Nobelstiftelsen, time is constantly reshaping itself, stretching and bending like the figures of rubber-sheet geometry. Present recapitulates past, anticipates future. Small wonder our contemplation of this hallowed place led us to imagine that the dust motes dancing in the light as Dr. Roger Sperry received his Nobel Prize were the same ones that danced when Albert Einstein won his. Nor is it surprising that Sperry's being honored in 1981 should evoke and recapitulate an earlier moment, in 1963, when Sir John Eccles was introduced as one of that year's Laureates in Physiology or Medicine by Professor Ragnar Granit:

Sir John Eccles

PROFESSOR GRANIT

Your Majesties, Your Royal Highnesses, Ladies and Gentlemen,

Sir John Eccles' discoveries concern the electrical changes which the nerve impulses elicit when they reach another nerve cell. In this experiment the microelectrode, with a tip of less than $\frac{1}{1,000}$ mm, is placed, for instance, in a so-called motoneurone in the spinal cord. These motor cells have a diameter between 40 and 60 thousandth of a mm. The arriving impulse produces excitation or inhibition in the motor cell, because the terminals of the nerve fibre are connected to excitatory or inhibitory chemical mechanisms at the cell membrane. These are called synaptic mechanisms because the points of contact are known as synapses. There are two kinds of synapses, one excitatory, the other inhibitory. If the arriving impulse is connected to excitatory synapses the response of the cell is *yes,* i.e., excitability increases. *Vice versa* the inhibitory synapses make the cell respond with a *no,* a diminution of excitability. Eccles has shown how excitation and inhibition are expressed by changes of membrane potential.

When the response is sufficiently strong to cause excitation, the membrane potential decreases until a value is reached at which the cell fires off an impulse. This impulse travels through the nerve fibre of the cell and causes contraction in a muscle. Obviously a cell may also send impulses to another cell at whose membrane the synaptic processes repeat themselves with a plus or minus sign, as the case may be.

A cell engaged in activity may be influenced by impulses reaching inhibitory synapses. In this case the membrane potential increases and, as a consequence, the impulse discharge is inhibited. Thus excitation and inhibition correspond to ionic currents which push the membrane potential in opposite directions.

The nerve cells are provided with thousands of synapses which correspond to terminals of fibres originating in sense organs or other nerve cells. The sum total of synaptic processes determines the state of balance between excitation and inhibition in which the integrated messages of nerve cells find expression and the code of impulses its interpretation.

(Turning to Sir John, Professor Granit concludes his introduction in the traditional manner:)

It is a privilege and pleasure for me to convey to you the warmest congratulations of the Nobel Assembly of Karolinska Institutet and to invite you to receive your Nobel Prize from the hands of His Majesty the King.

COMMENTARY

Let us consider Sir John a moment. He looks so self-possessed, unflappable, yet he is as passionately convinced as Sperry is that mind can't be reduced to neural electronics. Eccles is the one who showed that mental acts of intention *initiate* the burst of discharges in a nerve's brain cell. He has tried to re-enfranchise the human mind, to get science to recognize thinking as a more comprehensive human activity than the mere operation of neural mechanisms. One more thing Eccles's work shows is that voluntary decisions reached through mental processes are really possible for us. We are not mere automatons pushed about by the brain's chemistry. We possess a genuine power of choice, and that implies, further, a moral responsibility. Our world can be really changed for good or ill by our choices. Sir John is one of those scientists courageous enough to point out the social and ethical implications of his research. Like many other Nobel Laureates, Eccles has achieved his reputation by a willingness to transcend sacred convention, to pursue controversial ideas fearlessly. For him, the world and human life are not merely projects to be analyzed, but wonders to be contemplated and celebrated.

One thinks of Sir John Eccles as belonging to that long tradition of British culture which combines sturdy dedication to empirical research with an equally strong poetic sensibility. In his life and work one detects the pattern of those ancient Anglo-Saxon scholars like Bede the Venerable, whose contemplation of nature was based on minute observation of the world, its swarming diversity of life, its climates and peoples, and its history. Bede found a way to write of science that bore the happy marks of a humane literacy, and this tradition continues in works such as Sir John's recent *The Wonder of Being Human.*

    ▫

Another representative of the Anglo-Saxon tradition which blends poetic insight with scientific rigor is Dr. Brian Josephson. His pioneering research into tunneling phenomena in solids was rewarded with the Nobel Prize in Physics in 1973. Dr. Josephson, the young scholar from Cambridge, was introduced by Professor Stig Lundquist:

Brian Josephson

PROFESSOR LUNDQUIST

Your Majesties, Your Royal Highnesses, Ladies and Gentlemen,

The 1973 Nobel Prize for physics had been awarded to Drs. Leo Esaki, Ivar Giaver and Brian Josephson for their discoveries of tunneling phenomena in solids.

The tunneling phenomena belong to the most direct consequences of the laws of modern physics and have no analogy in classical mechanics. Elementary particles such as electrons cannot be treated as classical particles but show both wave and particle properties. An electron and its motion can be described by a superposition of simple waves, which forms a wave packet with a finite extension in space. The waves can penetrate a thin barrier, which would be a forbidden region if we treat the electron as a classical particle. The term tunneling refers to this wave-like property; the particle "tunnels" through the forbidden region. In order to get a notion of this kind of phenomenon let us assume that you are throwing balls against a wall. In general the ball bounces back but occasionally the ball disappears straight through the wall. In principle this could happen, but the probability for such an event is negligibly small.

On the atomic level, on the other hand, tunneling is a rather common phenomenon. Let us instead of balls consider electrons in a metal moving with high velocities towards a forbidden region, for example a thin insulating barrier. In this case we cannot neglect the probability of tunneling. A certain fraction of the electrons will penetrate the barrier by tunneling and we may obtain a weak tunnel current through the barrier.

Brian Josephson's theoretical discoveries showed how one can influence supercurrents by applying electric and magnetic fields and thereby control, study and exploit quantum phenomena on a macroscopic scale. His discoveries have led to the development of an entirely new method called quantum interferometry. This method has led to the development of a rich variety of instruments of extraordinary sensitivity and precision with application in wide areas of science and technology.

The applications of solid state tunneling already cover a wide range. Many devices based on tunneling are now used in electronics. The new quantum interferometry has already been used in such different applications as measurements to study the electromagnetic field around the heart or brain.

(Professor Lundquist addresses Dr. Josephson, inviting him to receive his prize from King Gustav:)

On behalf of the Royal Academy of Sciences I wish to express our admiration and convey to you our warmest congratulations. I now ask you to proceed to receive your prize from the hands of His Majesty the King.

COMMENTARY

Josephson's fascination with tunneling phenomena was rooted in his observation that electrons sometimes behave "mischievously" by penetrating barriers which, given earlier scientific theories, ought to be unbreachable. To notice some mischief among electrons seems in keeping with Josephson's character, for behind his dark-rimmed glasses are eyes that fairly snap with elfin merriment. Combining meticulous attention to detail with an almost childlike capacity for amusement has, indeed, been characteristic of many a British scholar. One thinks, for example, of J.R.R. Tolkien, whose career as a learned medievalist at Oxford seemed to blend so naturally with his life as a novelist who celebrated the mystery and mirth of Middle Earth in his beloved trilogy *The Lord of the Rings.* It is not too far-fetched to imagine that tucked behind the chart of periodic tables in Josephson's laboratory at Cambridge is a well-worn map of Middle Earth.

As Josephson leaves the rostrum in the hall of the Nobelstiftelsen to rejoin his beautiful wife and young children, one catches a glimpse of the hobbit in him. He seems to spring forward as he walks, thus giving the impression of a young athlete who might be more comfortable on the squash court than surrounded by solemn luminaries from the world of science. But Josephson's youthful appearance does not detract from his stature in the scientific community. Like Laureates Sperry and Eccles, Josephson is willing to risk raising questions other scientists may regard as controversial or heterodox. His proposal that "God" be included as an integral factor in serious scientific discussion has raised many an eyebrow, much as Tolkien's creation of Middle Earth astonished a few of his more conservative colleagues.

◆§   §◆

But let us abandon Middle Earth for the moment and turn our attention to Ilya Prigogine. Dr. Prigogine was born in Russia, but emigrated with his parents to Belgium. His keen interest in art and literature, combined with his originality as a research chemist in the area of thermodynamics, have earned him a unique place in the history of science. At the Nobel Prize ceremonies of 1977, Prigogine was introduced by Professor Stig Claesson:

(Photo of Ilya Prigogine, overleaf)

Ilya Prigogine

PROFESSOR CLAESSON

Your Majesties, Your Royal Highnesses, Ladies and Gentlemen,

The discoveries for which Ilya Prigogine has been awarded this year's Nobel Prize for Chemistry come within the field of thermodynamics, which represents one of the most sophisticated branches of scientific theory and is of enormous practical relevance.

The history of thermodynamics dates back to the early years of the nineteenth century. However, it was subject to certain limitations. For the most part it could only deal with reversible processes, that is, processes occurring via states of equilibrium.

Prigogine's great contribution lies in his successful development of a satisfactory theory of non-linear thermodynamics in states which are far removed from equilibrium. In doing so he has discovered phenomena and structures of completely new and completely unexpected types, with the result that this generalized, nonlinear and irreversible thermodynamics has already been given surprising applications in a wide variety of fields.

Prigogine has been particularly captivated by the problem of explaining how ordered structures—biological systems, for example—can develop from disorder. The classical principles of equilibrium in thermodynamics still show that linear systems close to equilibrium always develop into states of disorder which are stable to perturbations and cannot explain the occurrence of ordered structures.

Prigogine and his assistants chose instead to study systems which follow non-linear kinetic laws and which, moreover, are in contact with their surroundings so that energy exchange can take place—open systems, in other words. If these systems are driven far from equilibrium, a completely different situation results. New systems can then be formed which display order in both time and space and which are stable to perturbations. Prigogine has called these systems dissipative systems, because they are formed and maintained by the dissipative processes which take place because of the exchange of energy between the system and its environment and because they disappear if that exchange ceases. They may be said to live in symbiosis with their environment.

The method which Prigogine has used to study the stability of dissipative structures is of very great general interest. It makes it possible to study the most varied problems, such as city traffic problems, the stability of insect communities, the development of ordered biological structures and the growth of cancer cells, to mention but a few examples.

Thus Prigogine's researches into irreversible thermodynamics have fundamentally transformed and revitalized science, given it a new relevance and created theories to bridge the gaps between chemical, biological and social scientific fields of inquiry. His works are also distinguished by an elegance and a lucidity which have earned him the epithet "the poet of thermodynamics."

(Professor Claesson concludes by speaking directly to Dr. Prigogine:)

Professor Prigogine, I have tried briefly to describe your great contribution to non-linear irreversible thermodynamics, and it is now my privilege and pleasure to convey to you the heartiest congratulations of the Swedish Royal Academy of Sciences and to ask you to receive your Nobel prize from the hands of His Majesty the King.

❧   ☙

SUMMING UP

Our contemplation of the Nobel Prize ceremonies in Stockholm, and our imagined encounters with Laureates past and present, are drawing to a close. The old hall of the Nobelstiftelsen pulses with the decades-old presence of important people and important ideas. As we listened to the short speeches honoring Roger Sperry, Sir John Eccles, Brian Josephson and Ilya Prigogine, we were struck by some of the heady research that is shaping the human future: the brain's hemispheres, neuronal events enfolded by mental intentions, the integrated messages of nerve cells, the life-like tunneling of mischievous electrons, the creative, irreversible arrow of time-becoming. Perhaps, as we listened, one of Bergman's angels, or Rilke's, or spectres from past Nobel ceremonies, drifted up into the huge thoughtful Swedish night.

Meanwhile, we prepare for a trip westward, a third of a world away, to the city of Dallas and its Isthmus Institute. Our journey from Stockholm to Dallas involves not merely a change of climate, but a change of internal time and geography as well. Built on a network of islands, peninsulas and waterways, Stockholm is a Venice of the North, enriched by many centuries of art, architecture and cultural life. In contrast, Dallas is a parvenu among the world's great metropolitan centers. Less than one hundred and fifty years ago, there was no permanent human settlement at all on the banks of Texas's Trinity River. Today, the river is hardly visible, for dozens of skyscraping glass towers slice through the horizon to reflect the hot sun of the Southwest. Where no one lived in the 1830s there has risen a sprawling metroplex of three and a half million people.

Dallas's rapid growth and economic expansion are perhaps less remarkable, however, than its impact on the imagination. Leaving aside such fictions as South Fork and the oil-loving Ewing family, one sees in Dallas the powerful symbolic attraction of a human community organizing itself on the edge of a desert. Dallas's geographical location provides a constant tension between the human manipulation of space and the potent, disorganized vastness of a desert-frontier, consuming, chaotic, waterless and thus life-threatening. Not far away, in Houston, another frontier grips one's consciousness; the NASA laboratories and the Lyndon B. Johnson Space Center trigger flights of imagination into the future toward that "diminishing air past. . . . all the light diffusions and bombardments" of which A. R. Ammons reminded us in the poem which opened this "Prelude."

Our pilgrimage from Stockholm to Dallas thus draws us from symbol to symbol. In the old hall of the Nobelstiftelsen, time is condensed and recapitulated, while on the shimmering surfaces of Dallas's glass towers, time rushes irreversibly into the future. This symbolic movement seems focused in the career of Nobel Laureate Ilya Prigogine, who divides his time between the ancient European capital of Brussels and the modern city of Austin, home of the University of Texas.

The Isthmus Institute of Dallas, which sponsored the Nobel Conversations in this book, represents a constellation of persons who combine reverence for wisdom already acquired with a driving concern for the human future. Isthmus brings scholars and philanthropists together in an effort to sponsor events of major cultural import for citizens of the American Southwest. Its support is drawn from the Meadows Foundation, the Texas Council of the Humanities, the Communities Foundation of Texas, as well as from business and industrial leaders, academic and professional people, artists and students.

The primary goal of the Isthmus Institute closely parallels that of the Nobel Foundation. Both seek to honor those gifted persons in the sciences and humanities who use their knowledge to serve the highest interests of humankind. Both these philanthropies unite in a common conviction: that the human future is, quite simply, the humans of the future. For as A. R. Ammons notes in his "Hymn," our human dignity dwells precisely in that dual capacity to "leave the earth and go on out. . . . past the blackest noctilucent clouds" and to "stay with the earth. . . . trusting the microvilli sporangia."

         The day itself
Is simplified: a bowl of white,
Cold, a cold porcelain, low and round,
With nothing more than the carnations there.

Say even that this complete simplicity
Stripped one of all one's torments, concealed
The evilly compounded, vital I
And made it fresh in a world of white, . . .
Still one would want more, one would need more, . . .

There would still remain the never-resting mind,
So that one would want to escape, come back
To what had been so long composed.

From "The Poems of Our Climate,"
by Wallace Stevens

CONVERSATION ONE

ह✦

MIND ENFOLDS BRAIN:
THE MENTALIST REVOLUTION

ह✦

AN EXCHANGE BETWEEN
ROGER SPERRY AND SIR JOHN ECCLES

ABOUT CONVERSATION ONE

The "Prelude" which opened this book found us in
Stockholm, amid the glittering social and intellectual atmosphere of the Nobel Prize ceremonies. Attention was
focused on the personalities and work of Roger Sperry, Sir
John Eccles, Ilya Prigogine and Brian Josephson. At this
point we are prepared to begin listening to the Laureates
themselves. Traveling across the Atlantic, then halfway

across the continental United States, we arrive in Dallas, Texas, where the Isthmus Institute conversations were held in November, 1982.

Though Wallace Stevens was neither a Texan nor a winner of the Nobel Prize, he is generally regarded as one of the finest American poets of the twentieth century, and lines from his "The Poems of Our Climate" have been chosen to introduce Conversation One. Throughout his career, Stevens was fascinated by the relation between inner and outer worlds, by the interplay between the "factual" world of nature "out there" and the interior world of human thought and reason, idea and imagination. The never-resting mind, Stevens insisted, hotly rebels against mere fact, even facts so lovely as clear water and a shining bowl of pink and white carnations. The mind's reach is long, its embrace large, its hunger hard to satisfy. As Stevens remarks in a later passage of the same poem, quoted further on in Conversation One, "The imperfect is our paradise," and our bittersweet delight as thinking humans thus "lies in flawed words and stubborn sounds."

The power of the never-resting mind, as it acts through mental intentions, thinking and decisions that enfold and supersede the purely material functions of the brain, forms the central theme of this first exchange between Roger Sperry and Sir John Eccles. What these scientists have to say, though sometimes couched in technical language, is ultimately neither arcane nor unintelligible. The evolving universe they describe is none other than the one in which all of us are embedded and for which all of us are responsible.

We find ourselves in a universe that is ever expanding, changing, becoming more complex, more mysterious, richer in meaning. Time was, not so long ago, when some scientists saw the world as a clock winding down and running out of energy. The universe appeared to have yielded most of its secrets. It remained merely to demonstrate mathematically how all reality is governed by immutable laws whose exact application can be determined by science with infallible accuracy.

Few of us feel comfortable with that view of the universe or that kind of science. Our experience tells a different story. Change, becoming, and time steadily unfolding are the realities common to our everyday experience. From the moment we choose what to eat for breakfast to the moment we switch off the late-night drone of Ted Koppel, our days are spent on the edge of time, where things are always "becoming," where choices and intentions, however trivial, enfold the activities of our brains.

So when Sperry and Eccles insist in the following conversation that mind is in charge of brain, we spontaneously recognize their conviction as something we've always known or at least suspected. What grips us as we listen to these men is not only the elegance of their demonstrations, nor the sheerly rational force of their arguments, but their everydayness. And like most of our quotidian experiences, the ones described by these scientists supersede "explanation:" they are better grasped at what might be called the level of intuitive "rightness." We find ourselves agreeing with Sperry and Eccles because what they say seems "right." We listen to these scientists, therefore, in much the same way we listen to convincing poetry and music. When Emily Dickinson writes

> I felt a Funeral, in my Brain,
> And Mourners to and fro
> Kept treading—treading—

we know exactly what she's talking about, even if we can't explain it. When Allen Ginsberg howls at us, flinging out venomous lines like

> I saw the best minds of my generation destroyed
> by madness,
>   starving hysterical naked, . . .
> angel-headed hipsters . . . ,

we instinctively recognize and feel the painful confusion, even if we recoil from beatniks, drug-heads and hippies. Listening to Beethoven's Ninth Symphony discloses something fresh and joyful about being human, even if at the strictly rational level we know he wrote the work when he was deaf, aging and possibly syphilitic.

If you thus find yourself reacting to what these scientists say in much the same way you react to literature and the arts, you're on the right track. It is at least worth asking whether or not there is such a vast conflict between the anonymous sculptor who released coiling vitality from stone in the Belvedere Torso and the well-known physicist who sought to release energy from atoms. Sculptor and physicist are both artists; both construe the world itself as art, and the media they employ may be more similar than at first meets the eye. As the universe continues its becoming, the frontiers of both science and art further unfold and expand.

The first Isthmus Conversation thus leads us not only from Stockholm to Dallas, but further still, to the common creative springs of the humanities, arts and sciences. Throughout this and the following Conversations, the Laureates' names are identified in capitals whenever they are speaking. In addition, there is another voice in each Conversation, editorially condensed and extrapolated from comments made by guests at the Isthmus Institute proceedings. This editorial voice is heard in the introduction and in the "Summing Up" of each Conversation, where it is set off from the rest of the text by printers' ornaments. Within the Conversations themselves, this voice is labelled "Commentary" and its contributions are printed in *italics*.

A common interest unites the speakers in Conversation One. Eccles and Sperry both seek to rehabilitate the distinctive position of mind, often ignored by scientific materialists or reduced to the status of a negligible "effect" produced after the brain has done all the important electrochemical work. The research of Sperry and Eccles re-enfranchises those mental qualities and activities which experience tells us are quintessentially human: voluntary choice, intention, personal freedom, and responsibility.

And now, the historic Nobel Prize Conversations held at Dallas's Isthmus Institute in November, 1982. Dr. Roger Sperry is the first speaker.

᧪   ᨀ

DR. SPERRY

The former scope of science, its limitations, world perspective, views of human nature, and its societal role as an intellectual, cultural and moral force are all undergoing profound change. Where there used to be a chasm and irreconcilable conflict between the scientific and the traditional humanistic views of man and the world, we now perceive a continuum. A unifying new interpretative framework emerges with far-reaching impact not only for science but for those ultimate value-belief guidelines by which mankind has tried to live and find meaning.

The revisions in science I refer to have advanced farthest, and are most clearly manifest in the mind-brain and behavioral sciences, in what has come to be called the "consciousness" or "mentalist" revolution of the 1970s. A broad shift of conceptual framework or scientific paradigm is involved, a shift in psychology from objective behaviorism to a more subjective cognitivism, from the old reductive materialism to a new more holistic mentalism. The outcome today brings revised concepts of brain and consciousness, of free-will and the inner self, and of the make-up of human nature in general.

Here I would like to summarize some of the research that has led me to revise my own understanding of the human mind. The relation between the brain's left hemisphere (logical, analytic) and its right hemisphere (long believed mute and lacking in a sense of "self") was the thing that especially intrigued me. Earlier contentions that the right hemisphere is not even conscious had largely given way by the mid-seventies to an intermediate position conceding that the mute hemisphere may be conscious at some lower elemental levels, but claiming that it lacks the higher, reflective, self-conscious kind of inner awareness that is special to the human mind and is needed, so it is said, to qualify the right conscious system as a "self" or "person."

Accordingly we undertook to test the right hemisphere more specifically for the presence of self-recognition and related forms of self and social awareness. With perception of pictorial stimuli confined to one hemisphere by the scleral contact lens occluder developed by Eran Zaidel, the subject merely had to point to select items in a multiple choice array in answer to various kinds of leading questions regarding his or her knowledge and feelings concerning the content of the pictures.

The results revealed that the disconnected right hemisphere readily recognizes and identifies him or herself among a choice array of portrait photos, and in doing so, generates appropriate emotional reactions and displays a good sense of humor requiring subtle social evaluations. Similar findings were obtained for pictures of the immediate family, relatives, acquaintances, pets, personal belongings, familiar scenes and also political, historical and religious figures, as well as television and screen personalities. The relatively inaccessible inner world of the nonspeaking hemisphere was thus found to be surprisingly well developed. Results to date suggest the presence of a normal and well developed sense of self and personal relations along with a surprising knowledgeability in general.

Similar procedures were used to explore for a sense of time in the right hemisphere and the presence of concern for the future with thus far no evidence of abnormal deficit. The nonvocal hemisphere appears to be quite cognisant of the person's daily and weekly schedules, the calendar, seasons, and important dates of the year. The right hemisphere also makes appropriate discriminations that show concern with regard to the thought of possible future accidents and personal or family losses. The need for life, fire, and theft insurance, for example, seems to be properly appreciated by the extensively tested mute hemisphere of these patients.

Unlike other aspects of cognitive function, emotions have never been readily confinable to one hemisphere. Emotional effects tend to spread rapidly to involve both hemispheres, apparently through crossed fiber systems in the undivided brain stem. In the tests for self-consciousness and social awareness it was found that even subtle shades of emotion or semantic connotations generated in the right hemisphere could be quite helpful to the left hemisphere in its efforts to guess the nature of a stimulus known only to the right hemisphere.

More structured components of cognitive processing were shown to be separable from emotional components.

Cognitive processing remained confined within the hemisphere in which it was generated, whereas emotional overtones leaked across to influence neural processing in the other hemisphere. The evidence of this separability is in itself significant in regard to questions of the organization of the neural mechanisms of cognition. Also, since the emotional component appears to be an eminently conscious property, the fact that it crosses at lower brainstem levels is of interest in reference to the structural basis of consciousness.

One of the more important things to come out of the split-brain work is a revised concept of the nature of consciousness and its fundamental relation to brain processing. The key development here is a new causal interpretation that ascribes to inner experience an integral causal control role in brain function and behavior. In effect, and without resorting to dualist views, the mental forces and properties of the conscious mind are restored to the brain of objective science from which they had long been excluded on materialist-behaviorist principles.

COMMENTARY

This is surely fascinating, Dr. Sperry, yet it raises a number of questions. Our customary attitude toward the emotional components of brain processing is irritation or bemusement. Emotions have been regarded as unwelcome intruders that meddle with the brain's work of "pure cognition." But here, you imply that emotional effects, with all their finely-graded shading, contribute positively across both hemispheres of the brain to the work of cognitive processing. Wouldn't this mean that emotions are an intrinsic part of what we've always called "thinking"? And further, wouldn't this

require us to incorporate mental acts and intentions into our under-
standing of what the brain is and how it works? Indeed, wouldn't
your view result in giving the events of inner experience—ideas,
ideals, intentions, emotional tone—a certain priority over the
strictly neuronal operation of the brain?

DR. SPERRY

Yes, acceptance of the revised "causal view" carries important implications for science and for scientific views of
man and nature. Cognitive introspective psychology and
related cognitive science can no longer be ignored experimentally, or written off as "a science of epiphenomena,"
nor as something that must, in principle, reduce eventually
to neurophysiology. The events of inner experience, as
emergent properties of brain processes, become themselves
explanatory causal constructs in their own right, interacting
at their own level with their own laws and dynamics. The
whole world of inner experience (the world of the humanities) long rejected by twentieth century scientific materialism, thus becomes recognized and included within the domain of science.

Basic revisions in concepts of causality are involved here;
the whole, besides being "different from and greater than
the sum of the parts," also causally determines the fate of
the parts, without interfering with the physical or chemical
laws for the subentities at their own level. It follows that
physical science no longer perceives the world to be reducible to quantum mechanics or to any other unifying ultra
element or field force. The qualitative, holistic properties at
all different levels become causally real in their own form

and have to be included in the causal account. Quantum theory on these terms no longer replaces or subsumes classical mechanics but rather just supplements or complements.

Conscious or mental phenomena are dynamic, emergent, pattern (or configurational) properties of the living brain in action—a point accepted by many, including some of the more tough-minded brain researchers. Second, my argument goes a critical step further, and insists that these emergent pattern properties in the brain have causal control potency—just as they do elsewhere in the universe. And there we have the answer to the age-old enigma of consciousness.

To put it very simply, it becomes a question largely of who pushes whom around in the population of causal forces that occupy the cranium. There exists within the human cranium a whole world of diverse causal forces; what is more, there are forces within forces within forces, as in no other cubic half-foot of universe that we know. At the lowermost levels in this system are those local aggregates of subnuclear particles confined within the neutrons and protons of their respective atomic nuclei. These individuals, of course, don't have very much to say about what goes on in the affairs of the brain. Like the atomic nucleus and its associated electrons, the subnuclear and other atomic elements are "molecule-bound" for the most part, and get hauled and pushed around by the larger spatial and configurational forces of the whole molecule.

Similarly the molecular elements in the brain are themselves pretty well bound up, moved, and ordered about by the enveloping properties of the cells within which they are located. Along with their internal atomic and subnuclear parts, the brain molecules are obliged to submit to a course of activity in time and space that is determined very largely by the overall dynamic and spatial properties of the whole brain cell as an entity. Even the brain cells, however, with their long fibers and impulse conducting elements, do not have very much to say either about when or in what time pattern, for example, they are going to fire their messages. The firing orders come from a higher command.

The flow and the timing of impulse traffic through any cell, or nucleus of cells, in the brain is governed very largely by the overall encompassing properties of the whole cerebral circuit system, and also by the relationship of this system to other circuit systems. Even the circuit properties of the cerebral system as a whole, and the way in which these govern the flow pattern of impulse traffic throughout—that is, the circuit properties of the whole brain—may undergo radical and widespread changes with just the flick of a cerebral facilitatory "set." This set is a shifting pattern of central excitation that will open or prime one group of circuit pathways while at the same time closing, repressing, or inhibiting endless other circuit potentialities. Such changes of set are involved in a "shift of attention," "a turn of thought," "a change of feeling," or "a new insight," etc. In short, if one climbs upward through the chain of command within the brain, one finds at the very top those overall organizational forces and dynamic properties of the large patterns of cerebral excitation that constitute the mental or psychic phenomena.

COMMENTARY

Dr. Sperry, this exposition appears to approach research and knowledge from a non-reductionist point of view. That is, it does not subscribe to the traditional scientific method of understanding phenomena by reducing them to their component parts. In fact, it seems to question the reality of separable "component parts." Your work restores properly mental qualities and actions to human life. Further, you show that the mind is always "becoming" as it evolves through time to enfold rather than replace the activities of its sub-elements. The mind's enfolding power seems thus to extend to time as well. One recalls, in this connection, lines from another of Wallace Stevens's poems, "Sunday Morning,"

> There is not any haunt of prophecy,
> Nor any old chimera of the grave,
> Neither the golden underground, nor isle
> Melodious where spirits gat them home,
> Nor visionary south, nor cloudy palm
> Remote on heaven's hill, that has endured
> As April's green endures; or will endure
> Like her remembrance of awakened birds,
> Or her desire for June and evening, tipped
> By the consummation of the swallow's wings.

Through its dual power of memory and anticipation, mind enfolds time passing and time future. Thus, even when earth's satisfactions disappear—as the woman referred to in the last lines of the stanza knows they will—some things earthly will endure: April's green, in remembrance; June and evening, in anticipation. Perhaps there is a resemblance, Dr. Sperry, between Stevens's suggestion that mind is enfolded in time and your conviction that real mental acts supersede, though they don't replace, the brain's electro-chemical activity.

In any event, your scientific convictions are surely a testament to your intellectual courage. Your position here departs radically from the popular view that the human brain is basically a self-regulating computer, with mind little more than an after-effect produced to pacify philosophers and theologians. You deny that mind is pushed around by brain's activities at the subnuclear, molecular, cellular or circuitry levels. Mind exerts causal control—by choosing, commanding, willing—over the brain's functions.

SIR JOHN ECCLES

Let me approach Dr. Sperry's subject from a different angle. We have the indubitable experience that by thinking and willing we can control our actions if we so wish, such actions being called voluntary movements. The sequence can be expressed as motive (thinking), intention (willing), voluntary action. In bringing about movement voluntarily some brain events are initiated. There is the well-known crossed pyramidal tract from the motor cortex down the spinal cord to the nerve cells of the opposite side that cause the muscles to contract. The motor cortex is a narrow band of the cerebral cortex running over its convexity from the midline near the vertex. The left motor cortex controls the right side of the body and vice versa.

It might be thought that voluntary movement is so explained, but the reality is enormously more complicated and only partly understood. The pyramidal cells of the motor cortex discharge impulses down the pyramidal tract to motoneurons in the spinal cord that control movement. But this is only the last stage in the brain events concerned in voluntary movement. There are fundamental problems. In the initiation of a voluntary movement by a motive/intention, what are the sequences of events being set up in the brain?

A remarkable series of experiments in the last few years have transformed our understanding of the cerebral events concerned with the initiation of a voluntary movement. It can now be stated that the first brain reactions caused by the intention to move are in nerve cells of the supplementary motor area (S M A). It is right at the top of the brain. This area was recognized by the renowned neurosurgeon, Wilder Penfield, when he was stimulating the exposed human brain in the search for epileptic "foci" (regions of aberrant activity associated with epileptic seizures). Stimulation of this area did not cause sharply localized responses for the motor cortex. Instead there were writhing or adversive movements of large parts of the torso and limbs, even of the same side, and also incoherent vocalizations. So this area was neglected for decades as it did not seem to have an interesting function. Now, from this Cinderella status, the SMA has been advanced to a role of highest interest.

There is strong support for the hypothesis that the SMA is the sole recipient area of the brain for mental intentions that lead to voluntary movements. This is a most important improvement on the concept that the mental act of intention to move was widely dispersed in its action on the brain.

Such a sharp focusing lends precision to our attempts to define the manner in which some particular voluntary action is brought about. The concept of motor programs is important in this enterprise. Instead of just waving our arms or indulging in some other crude movement to display intention leading to action, we have to recognize the complexity of the muscle actions in bringing about any skilled and learned movement. It can be as commonplace as extending one's hand to pick up a cup, and, in an elegant and smooth action, to bring the cup to one's lips for a drink, thereafter placing the cup back in its saucer. A most complex series of movements has been accomplished, each of which can be reduced to some constituent motor programs; putting one's hand to the cup; securely grasping the handle; smoothly lifting the cup; bringing it correctly to the lips; drinking, which is a whole series of lip, tongue, pharyngeal and swallowing movements; and finally the return of the cup. Thus there is a whole interlocking series of movements involving contraction of a large number of muscles, all nicely graded and sequenced. It is convenient to describe this complex movement as being composed of a harmony of elemental motor programs.

Let us now consider how our intention to enact such a voluntary movement can be related to the role of the SMA as the mediator between the mental act of intention and the assemblage of motor programs involved in the voluntary movement. We must first postulate that the mental intention acts on the SMA in a highly selective way, and that the SMA contains, as it were, an inventory of all the learned motor programs. This immense stored repertoire of the learned motor programs of a lifetime could not be stored in the SMA, which is a quite limited area of the cerebral cortex with perhaps 50 million neurons and 15,000 modules on each side. All that is necessary is for the SMA to contain the inventory of the motor programs; an inventory which comprises addresses to the storage banks of the motor programs. The SMA is known to have major lines of communication to the presumed storage sites in the cerebral cortex (particularly the premotor cortex and in the basal ganglia and the cerebellum). By radio-tracer techniques these areas have been shown to be called into action in voluntary movements, and many nerve cells in these circuits have been shown to be active before the discharge of the motor cortical cells.

Thus we have in outline a hypothesis of how the mental act of intention can, by action through the SMA, bring about the desired movement.

Let us now return to the performance of the nerve cells of the SMA in a voluntary act. A single nerve cell under observation will be firing at the usual slow and irregular frequency, the background discharge. Then there is a sharp increase of the frequency of firing and in a little more than a fifth of a second the movement starts. We could in fact predict the onset of the movement from the observed discharges of the nerve cell. It is important to recognize that this burst of discharge of the observed SMA cell was not triggered by some other nerve cell of the SMA or elsewhere in the brain. The first discharges occur in the SMA.

So we have here an irrefutable demonstration that a mental act of intention initiates the burst of discharges of a nerve cell. Furthermore, when hundreds of SMA nerve cells are observed, only some are activated at this early stage. Others come later; others may fire in two successive bursts; still others are even silenced. And these specific types of response are repeatedly displayed by any nerve cell in relation to a particular voluntary act. Thus we have to postulate that the mental act of intention was being effective in a discriminating fashion. There was observed in fact a most complex pattern of neuronal responses of the SMA nerve cells. So the mental act of intention is exerted in a subtle discriminating fashion on the constituent nerve cells of the SMAs on each side.

COMMENTARY

So far, Sir John, you seem to be describing mental acts of intention as these occur in human subjects, as witness your example of the complex motor programs activated by the "simple" intention to lift a cup from a saucer and drink. Does this mean that the SMA's nerve cells respond to mental intentions only in the case of humans who are self-aware and self-reflective? In other words, does everything you've said about the SMA apply only to rational, reflective intelligence of the human sort?

SIR JOHN ECCLES

No. As a matter of fact, some of the important research on the SMA was done by Robert Porter and Corbie Brinkman, in whose studies a monkey had recording microelectrodes surgically implanted in the SMA. After recovery, the monkey initiated voluntary movements by pulling a lever in a self-paced manner with either hand in order to obtain a food-reward. It was found that with this voluntary act many of the nerve cells of the SMA began to discharge well before the cells in the motor cortex of the brain.

Even more interesting facts have been learned from further research. A particular voluntary intention—for example, the monkey's intention to pull the lever—is conveyed in the form of a discriminating code of the SMA neurons. This code must have a spatio-temporal pattern, for the action proceeds in space and time. We have entered into a field of discourse where discriminating mental events have highly selective actions on the SMA neurons. Presumably the mental influences are exerted by a code in a graded and varied manner and are subject to feedback influences from the activated SMA neurons. The frontier between mental events and neural events must be traversable in both directions.

It is evident that we have embarked upon a temerarious field of speculation. But the fact remains that skilled and learned movements are carried out at will; that immensely complicated neural machinery is necessary for any such act; and that the mental influences must work in a coded manner on the SMA neurons and generate corresponding codes of spatio-temporal patterns in the discharges of the SMA neurons. Each such pattern presumably is an inventory of motor programs with the addresses for transmitting the codes so as to institute the activities of these motor programs.

How can the mental act of intention activate across the mind-brain frontier those particular SMA neurons in the appropriate code for activating the motor programs that bring about intended voluntary movements? The answer is that, despite the so-called "insuperable" difficulty of having a non-material mind act on a material brain, it has been demonstrated to occur by a mental intention—no doubt to the great discomfiture of all materialists and physicalists.

COMMENTARY

What you have just said, Sir John, is quite astonishing, especially for those of us who may have been trained in the reductionist mode to believe that "mental activity" is nothing more than biochemical response and electrical impulse. It is rather difficult to shift gears and claim—as you have just done—that a non-material mind acts on a material brain. Until quite recently, science assumed that to attribute to non-material forces such as mental intentions any kind of "causal" potency or control is to lapse into primitive mysticism or vague religious feeling. What you are proposing here is certainly not mysticism in the usual sense of that word. It resembles more closely, perhaps, the kind of insight we find in Walt Whitman's poetry.

Whitman sensed that the activities of all life forms, from the quail whistling in the woods to the shark's fin cutting through dark water, are related to each other in a manner more sophisticated, and indeed more "intelligent," than we are usually inclined to believe. Vital forces, not reducible to mere material fact, work throughout the universe to constitute what Whitman called "the exquisite scheme" of creation. What we often dismiss as mere "nature" or "the natural" is in fact a subtle, complex web of life that is no less mysterious simply because it surrounds us everywhere. On the contrary, what is most natural—what is everywhere and touches life daily—fascinates and frightens us the most. Death, for example, or the perilous coming forth of life:

Walking the path worn in the grass . . .
Where the quail is whistling betwixt the woods
  and the wheat-lot,
Where the bat flies in the Seventh-month eve . . .
Where the cattle stand and shake away flies . . .
Where the human heart beats with terrible throes . . .
Where the pear-shaped balloon is floating aloft . . .
Where the life car is drawn on the slip noose . . .
Where the she-whale swims with her calf . . .
Where the fin of the shark cuts like a black chip
  out of the water . . .

Whitman postulated a kind of intelligent network, a mutually recognized, natural vital force that binds all forms of life together in the evolving universe. Perhaps his postulate reflects, Sir John, some of the complexity you note in the interactions between mental intentions, the role of the SMA and the electro-chemical activity of the brain.

In any event it is clear that both you and Dr. Sperry are upholding a "mentalist revolution" in science. Strictly orthodox materialists may doubt such a revolution and label it an atavistic throwback to "prescientific" perceptions of nature which believed that non-material reality could act on the material. But in fact, both of you have reached your conclusions through the rigorous discipline of the laboratory. If you are persuaded that mental realities initiate and direct biochemical reactions in the brain, it is scientific experimentation, not philosophical speculation, that has convinced you. And if your work resonates with the "exquisite schemes" of Whitman and the consummation of a swallow's wings in Stevens, perhaps that is because, in our day, scientific and humanistic modes of inquiry are converging in unexpected ways.

Others, as well, have suspected this convergence. Pierre Teilhard de Chardin, a Roman Catholic priest and palaeontologist whose works were banned by the Church during his lifetime, wrote of the universe evolving toward a condition in which human consciousness and mental activity, empowered by freedom and love, develop ever higher levels of integration. Such integration, Chardin thought, would eventually embrace all aspects of the universe, including the inanimate realities of our biosphere. He thus suspected a connectedness among all things in the universe, and guessed that human mind plays a decisive role in the interplay between brain dynamics and inner mental experience.

DR. SPERRY

Permit me to draw attention to some rather familiar experiences of the distinction between brain dynamics and inner mental experience. For simplicity, consider an elemental subjective sensation—and for reasons that will become evident let us use the sensation of pain instead of philosophy's old favorite, the color red. More specifically, make it pain in the wrist and fingers of the left hand of an arm that was amputated above the elbow some months previously. Suffering caused by pain localized in a phantom limb is no easier to bear than if the limb were still there. It is easier, however, with the example, to infer where our conscious awareness must reside.

With regard to this conscious sensation of pain, the contention is that any groans it may evoke—and any other response measures the patient may take as a result of the pain sensation—are indeed caused, not by the biophysics, chemistry, or physiology of the cerebral nerve impulses as such, but by the pain quality, the pain property, per se. This brings us to the real crux of the argument. Nerve excitations are just as common to pleasure, of course, as to pain, or any other sensation. What is critical is the unique patterning of cerebral excitation that produces pain instead of something else. It is the overall functional property of this pain pattern that is critical in the causal sequence of brain affairs. This pattern has a dynamic entity, the qualitative effect of which must be conceived functionally and operationally, and in terms of its impact on a living, unanesthetized cerebral system. This overall pattern effect in brain dynamics is the pain quality of inner experience.

Above simple pain and other elemental sensations in brain dynamics, we find, of course, the more complex but equally potent forces of perception, emotion, reason, belief, insight, judgment and cognition. In the onward flow of conscious brain states, one state calling up the next, these are the kinds of dynamic entities that call the plays.

It is exactly these encompassing mental forces that direct and govern the inner flow patterns of impulse traffic, including their physiological, electro-chemical, atomic, subatomic, and subnuclear details. It is important to remember in this connection that all of the simpler, more primitive, elemental forces remain present and operative; none has been canceled. These lower-level forces and properties, however, have been superseded in successive steps, encompassed or enveloped as it were, by those forces of increasingly complex organizational entities. For the transmission of nerve impulses, all of the usual electrical, chemical, and physiological laws apply, of course, at the level of cell, fiber and synaptic junction. Proper function in the uppermost levels depends to a large extent upon normal operation at the subsidiary levels. It is a special characteristic of these larger functional patterns in the brain, however, that they have a coherence and organization that enables them to carry on orderly function in the presence of considerable disruptive damage in the lower-level components.

Near the apex of this compound command system in the brain we find ideas. In the brain model proposed here, the causal potency of an idea, or an ideal, becomes just as real as that of a molecule, a cell, or a nerve impulse. Ideas cause ideas and help evolve new ideas. They interact with each other and with other mental forces in the same brain, in neighboring brains, and in distant, foreign brains. And they also interact with real consequence upon the external surroundings to produce *in toto* an explosive advance in evolution on this globe far beyond anything known before, including the emergence of the living cell.

COMMENTARY

When Dr. Sperry speaks of the "causal potency of an idea," one is strongly reminded once more of Wallace Stevens, for whom the ideas of the "never-resting mind" were surely as powerfully real as the white bowl and pink carnations of the poem which opened this Conversation. For Stevens, indeed, the potent ideas of the mind, interacting with other ideas and with the "facts out there," constitute both the grandeur and the misery of the human situation, for they lead us to prefer the complex to the simple, the composed to the plain, the accidents of imperfection to the dull sameness of the perfect. And thus he remarked, toward the end of "The Poems of Our Climate,"

There would still remain the never-resting mind,
So that one would want to escape, come back
To what had been so long composed.
The imperfect is our paradise.
Note that, in this bitterness, delight,
Since the imperfect is so hot in us,
Lies in flawed words and stubborn sounds.

*The evolutionary advance that results from ideas causing ideas
and helping evolve new ideas is also reflected in Stevens's "the
imperfect is our paradise." The perfect world of cold white porcelain
and still water is not the septic, teeming world from which life
emerges. Stevens believed that the restless mind acts with real effect
upon its environment, and this seems remarkably close to Dr.
Sperry's comment that ideas "interact with real consequence upon
the external surroundings to produce . . . an explosive advance in
evolution." From the human brain there have emerged those large,
dynamic patterns of coherence and organization—among them
such realities as perception, reason, insight and cognition—which
now permit us not merely to be carried along by the process of
evolution, but to be conscious of that process and to direct or alter
its course. A distinctive difference between brain and mind, then,
is our emergent mental ability consciously to choose evolution.*

SIR JOHN ECCLES

Let me give some examples of the differences I find between brain and mind. As I indicated earlier in this Conversation, what may be called "mental intentions," with their
ability to initiate a burst of discharges in a nerve cell, are
not confined strictly to the human species. I alluded to the

work of Robert Porter and Corbie Brinkman, whose laboratory monkeys initiated voluntary movements by pulling levers to obtain food. It was found that with this voluntary act, performed by simians, many of the nerve cells of the SMA began to discharge well before the cells in the motor cortex and indeed before any other nerve cells of the brain, except for a small focus in the premotor cortex, which is just anterior to the motor cortex.

This laboratory work with monkeys has produced some surprising results. Since a complex voluntary movement is caused by many muscles contracting in sequence, it could be anticipated that only some of the SMA nerve cells would be concerned in the muscle contraction initiating the lever-pulling. But it is impressive that many of the sample of several hundred SMA nerve cells were firing about one-tenth to one-fifth of a second *before* the earliest discharge of the pyramidal cells down the spinal cord. An important finding is that the nerve cells of one SMA are activated whether the monkey chooses to use the right or the left hand in the lever-pull. This may relate to the crossed activity of the motor cortex.

Secondly, there is the research of Nils Lassen, Per Roland and their group in Copenhagen. For more than a decade advantage has been taken of the insertion of a canula into the internal carotid artery of patients in order to study the cerebral circulation (angiography). With a radio-tracer technique, radio-xenon is injected through this canula into the cerebral circulation and, with a battery of 254 radiation detectors assembled in a helmet over the scalp, the circulation of blood can be simultaneously measured from that number of areas in a mosaic map of the cerebral cortex. The activity of nerve cells is accurately signalled by an increase in the circulation of blood. So in this way, during the carrying out of a wide variety of actions, the activities of the nerve cells of the cerebral cortex are measured and, by an exquisite technique, are immediately converted into a map of the cerebral cortex, color-coded for percentages of activity falling above or below the resting background. It is a wonderful technique and gives most impressive results, but there are two disabilities. Firstly, it takes about forty seconds of exposure to secure one picture. Secondly, the grain of the picture mosaic is rather coarse, the units of the mosaic being about one square centimeter or one-sixth of a square inch.

In this research a voluntary movement is chosen so that the patient has to concentrate continuously. The task is practised by the patient before the test so that it is well carried out. In a particularly significant test—called the motor sequence test—the thumb has to touch in quick succession finger-1 twice, finger-2 once, finger-3 three times, finger-4 twice. A new sequence with minimal pause begins with finger-4 touched twice, the patient now performing the original sequence in reverse order; then in the initial order; then in reverse order, etc., throughout the test. These movements require continuous voluntary attention and never become automatic. The invariable finding is that there are highly significant increases in blood flow in the hand area of the contralateral motor cortex and in the adjacent sensory area, as would be expected, and also in the SMAs of both hemispheres. There could, of course, be no evidence as to the priority of the SMA activation. This evidence was, however, obtained by a remarkable variant of the experiment that is called internal programming. The subject had to carry out the same motor sequence test but mentally without any movement whatsoever. This motor silence was checked by electromyography, a technique whereby even extremely subtle muscular contractions can be detected. As would be expected, there was no trace of activity in the motor cortex or the adjacent sensory cortex, but, wonderful to relate, the SMAs of both sides were activated almost as much as for the movement sequences, whereas all other brain areas showed no significant activity. So it can be concluded that, in the intending of a movement, nerve cells in the SMA are the first to be called into action. In the so-called internal programming, the mental intention initiates in the SMA, and nowhere else, the activity appropriate for the voluntary movement. Yet, at the same time, the mental influence restrains this activity from spreading elsewhere through the brain and so causing the motor cortex discharge down the spinal cord with the ensuing voluntary movement.

This finding closely parallels that of Porter and Brinkman's research program on the nerve cells of the monkey's SMA, which in a voluntary movement, often fired well before the cells of the motor cortex and other areas of the brain. A complementary finding in the radio-xenon experiments was that, with such simple movements as finger wagging or continuous pressure on a spring by thumb and forefinger, no activity of the SMA accompanied the activity of the motor cortex and its associated sensory cortex.

Doubtless a brief initial activity of the SMA was required to initiate the movement, but thereafter voluntary intention gave place to automatism. During the continuous voluntary intention required for the motor sequence test it is impossible to carry out a meaningful and subtle conversation. In contrast this can be done during continuous finger wagging or the exertion of pressure once such action has become automatic. These radio-xenon tests for SMA activity thus distinguish between voluntary and automatic movements.

These are only two examples from among many that could be cited from contemporary scientific research. But they show, I believe, a number of important things about the relation between mind and brain. First, they show that mental intentions truly exist and that they initiate the burst of discharges in a nerve cell that leads to voluntary movement (such as pulling a lever or moving the position of a finger)—even in the case of non-human subjects like the monkeys of Porter and Brinkman's experiments. We must be careful, therefore, not to confuse the real existence of mental intentions with the distinctively human activities of self-reflective thinking and its expression in language. The world of mental realities is more comprehensive than the world of human thought and discourse. Secondly, research has revealed the critical role of the SMA as the mediator between mental acts of intention and the bundle of motor programs involved in any voluntary movement. Moreover, we have discovered that mental intentions act upon the SMA in a highly selective, discriminating manner. In a fashion which is not yet fully understood, mental intentions are able to activate across the mind-brain frontier those *particular* SMA neurons that are coded for initiating the specialized motor programs that cause voluntary movements. As I remarked earlier, this may present an "insuperable" difficulty for some scientists of materialist bent, but the fact remains, and is demonstrated by research, that non-material mind acts on material brain.

ᵔᶾ   ᶾᵔ

A SUMMING UP OF CONVERSATION ONE

One gets the impression that Sir John attributes to mental intentions a kind of independent existence whose origins cannot be explained simply by appealing to the neural, electrochemical activity of the brain. Though he admits that mental influences are "subject to feedback influences from the activated SMA neurons," he seems to deny that such mental realities are constituted or caused by brain activity. The question naturally arises: where do mental intentions come from, what is their source, their origin?

In attempting to answer this question, perhaps one can discern a difference between the position of Sir John and that of Dr. Sperry. As Sperry said earlier in this Conversation, conscious or mental phenomena are "dynamic, emergent, pattern (or configurational) properties of the living brain in action." This seems to imply that the source of mental intentions is the brain itself in living action—but that once these emergent mental properties appear, they have causal control potency over the "lower" activities of the brain at the subnuclear, nuclear and molecular levels. Mind emerges from brain, then takes charge as chief or director in the complex chain of command within the brain. In Sperry's view, there is no need to appeal to any source outside the living brain in order to explain the origin and existence of mental phenomena. In contrast Sir John's position seems to leave the question of origins open-ended.

If this first Conversation concludes with an unresolved question about the ultimate origins of mental phenomena, we should not be surprised. The new science represented by Nobel Laureates Sperry and Eccles does not pretend to exhaust the mysteries of the evolving universe; still less does it claim that the world can be adequately interpreted on the basis of immutable laws that are strictly material in nature. A more open-ended universe, one richer in mystery and potential, is envisioned here. Older mechanistic and deterministic views of nature are giving way to a vision of the universe characterized by fluctuation, randomness and spontaneous activity. A far more complex and pluralistic picture of the world and our place in it emerges.

Recognition of this complexity and pluralism has led us to include the likes of Wallace Stevens and Walt Whitman in this Conversation. Their "presence" at the Isthmus Institute proceedings in Dallas is neither whimsical nor decorative, for the shaping of the world as a work of art requires the convergent insight of both scientists and poets. Walt Whitman's hunch, for example, that all living beings are bonded together in a kinship of mutually recognized "intelligence," seems astonishingly close to modern science's view that "we are all cousins." As Dr. Carl Sagan has recently written,

> All life on earth is the same life. There are su-
> perficial differences which . . . seem important to
> us. But down deep at the heart of life, we are, all
> of us—redwoods and nematodes, viruses and ea-
> gles, slime molds and humans—almost identical.
> We are all expressions of the interaction of pro-
> teins and nucleic acids.

This common life-root has evolved a human species that is conscious of itself and capable of choosing, through mental ideas and intentions, its further evolution. Dr. Sperry was not exaggerating when he remarked that the interaction between ideas and external surroundings produces an advance in evolution "far beyond anything known before, including the emergence of the living cell." Nor was Sir John Eccles claiming too much when he insisted that the action of non-material mind on material brain has been not merely postulated but scientifically demonstrated.

This never-resting, non-material mind is not content with the relative simplicity of "facts," nor with the precise registrations of the brain's neuronal equipment. As Stevens intimates in "The Poems of Our Climate," the human mind, hot for imperfection, yearns to disturb the perfect, to defy the simple, to muddy the waters. Mind makes a difference so critical that it enfolds, though it does not replace, the brain's elegant electronics.

You are drawn, as early man was drawn,
by the enchantment of dusk and flame
    to the council fire
. . . we gathered on the rim of the world
and watched the last fires of sun flare
    on the summits
and the valleys fill with cool rivers of night.
Stone, and hoary trees, and the bodies of our companions
merged in translucent unity with the world
    of mountain and sky; . . . .
A spirit of unearthly beauty
moved in the darkness and spoke in terms of song. . . .
You were aware of the almost mystic peace
    that came over us all
. . . the faces . . . of calm revelation . . . of adolescence,
hushed and surprised by the world's sharp beauty.

From the writings
of Ansel Adams

CONVERSATION TWO

ॐ

MIND ENFOLDS MATTER:
EVOLUTION AND GOD

ॐ

AN EXCHANGE BETWEEN
ROGER SPERRY AND BRIAN JOSEPHSON

ABOUT CONVERSATION TWO

In Conversation One, Dr. Roger Sperry and Sir John Eccles, joined by the ghostly presences of Walt Whitman and Wallace Stevens, dialogued about the relation between mind and brain. Their conversation converged upon a conviction, commonly shared, that mind is in charge of brain. Motives, ideas, ideals and mental intentions possess a causal control potency which, activated, initiates that burst of discharges in nerve cells required for voluntary movements from the simple (blinking an eye to clear it of dirt) to the complex (playing Bach's Dorian Fugue on an eighteenth-century organ). Further, such ideas and intentions, though non-material, are just as real as molecules, cells and nerve impulses. Further still, ideas reach out to other ideas in the same brain and in brains far distant in space and time. Left unresolved in the Conversation was a question of origins: whence come these mental phenomena? Sperry sees them emerging from the activity of the living brain itself, while Eccles holds open the possibility that they derive from an "elsewhere," the exact identity of which is not clearly defined.

Whatever the position one may hold about its origins, mind, as we experience it, is clearly an invitation to reach out, to go beyond, to transgress the limits of here-and-now by venturing toward other minds and toward an "elsewhere" that may be revealed only to the human imagination or to the heart's urging. In a word, human mind signals transcendence and invites transgression: transcendence of the self, transgression of the artificial limits imposed by the world of immediate fact. Indeed, science itself could not proceed if it did not imagine that things are not always what they seem to be.

Not accidentally, Conversation Two in the Isthmus Institute series is introduced by a quotation from the writings of the late photographer Ansel Adams. Adams's art celebrates the intuition expressed in the last sentence of the previous paragraph: things are not always what they seem to be. For photography is not merely the mute record of objects seen, but a lively revelation of the artist's inner eye and a sudden exposure to what is often barely noticed or not seen at all. Viewers of Adams's work are well aware that it is filled with mysterious presence—a presence not merely of light and shadow, but of dark, unearthly beauty. To see Adams's renditions of Yosemite or of California's sequoias is to feel the first stirrings of creation itself; it is to be, as he says, "hushed and surprised by the world's sharp beauty."

The art of Ansel Adams raises questions of a sort similar to the ones that surfaced in the preceding conversation between Roger Sperry and John Eccles. Considering the relation between mind and brain, we were led to ask about the origins of mental phenomena. Similarly, the artistry of Ansel Adams leads us to wonder about the source of a photograph's power: are we looking at it or is it looking at us? The question is not so easily answered. Whence comes the light that continues to illuminate that photographed tree, that snowcapped mountain, that face? Light speeds on, after all, at more than 186,000 miles per second, yet the light that spread over Yosemite in the 1930s seems somehow held, as through an act of enchantment, in an Adams photograph.

Things are not necessarily what they seem to be. For a long time, materialist science derided the notion that a non-material reality—a mind, a mental force—could have any influence at all over the electrochemical behavior of the brain. But as Sperry and Eccles demonstrated in their Nobel-Prize-winning research, that view was simply wrong. What seemed true, wasn't. In Conversation Two, modern science invites us to stretch our minds further still. If, as Sperry said in Conversation One, the mental forces and properties of the conscious mind are restored to the brain of objective science—and if inner experience thus exerts a causal control role in brain function and behavior—can we, do we, need to make any further claim for mind? Is it too extravagant to postulate that mind or mental forces somehow control the structure and behavior of matter itself? And if such a postulate were true, how might we describe that mind? Would it be necessary or desirable to resort to a religious language which speaks of "God," of a supreme intelligence or a universal mind?

The following Conversation between Roger Sperry and Brian Josephson touches these and related questions by showing how mind and mental phenomena are related to the larger universe of matter, time and space. Underlying the entire discussion is an issue vital for the human present and the human future. Are there really any "forces" which direct and control the evolving course of our universe? And if there are, what is their nature? Are they benign or malevolent, mental, material or both?

In the recent past, scientists often dismissed these questions as belonging to the less exact investigations of philosophy or theology. This reductionist position, as understood technically by scientists, would claim that how one evaluates "mind" or non-material mental forces makes little difference. Reductionists further assume that all biological processes can be explained by the same physical laws which chemists and physicists use to describe the behavior of "inanimate" matter. Matter itself was described as an inert and passive "substance" governed by mechanistic rules.

Dr. Roger Sperry repudiates the reductionist position

and argues that to reduce all scientific explanation to the laws of quantum mechanics or wave-and-particle physics is to distort our experience of reality and, indeed, to outrage the procedures of science itself. The world is simply too complex, too messy and unfinished, to permit reductionist interpretation. Quite the contrary is true: the universe we know and inhabit stubbornly resists all efforts to explain its nature and activity by a single, neatly compact set of laws.

Perhaps a rather homespun image may be of use here. Think of our world as expandable luggage you're trying to pack for a long trip. No matter what the salespeople have said about its capacity to hold everything you'll need, no matter how many members of your family have sat on it so it can be shut and locked, some unruly items always escape to hang out shamelessly from the sides. Usually, the items are somewhat embarrassing ones—intimate apparel, stockings, or that "special magazine" you wanted to take in case the trip got boring. Expandable luggage just won't be ruled by our desire for order and neatness.

So our universe. Vital forces work throughout our world, and these often defy interpretation solely by means of the laws associated with quantum mechanics or thermodynamics. This does not mean that such "laws" have no force or that order is utterly lacking in the evolving universe. Indeed, one of the arguments advanced by Dr. Brian Josephson in Conversation Two points to the presence of an "implicate order," intelligent and pervasive, at work throughout the evident randomness and fluctuation of evolution. Dr. Sperry does not appeal to the postulate of an implicate order, but he is convinced that the complex living realities of our world require a level of interpretation that goes beyond the rules of quantum analysis. Sperry's remarks begin with a kind of story about what happens when intelligent and highly skilled people—like scientists—look for the right things in the wrong places.

◆§   §◆

DR. ROGER SPERRY

Early biologists hoped to find the secret to life in the form of special vital forces that distinguish the living from the non-living or the animate from the inanimate. When they started looking into living things, however, no special vital forces could be discovered. The longer, the harder and deeper they looked, the more firmly biologists became convinced that there are no such things in this world as special vital forces. Instead, we concluded that all living things are nothing but physico-chemical processes in different forms and degrees of complexity, and that all life can be explained, in principle, by the laws of physics and chemistry. The idea that there exist any distinct "vital" forces came to be known as the doctrine of "vitalism" and by the 1930s had already become a subject of scorn and derision among nearly all biologists and remains so to this day.

What happened is that we biologists had been searching in the wrong places. You don't look for vital forces among atoms and molecules; you look instead among living things, i.e. among living cells and organisms acting and interacting as entities. You look, for example, among animals responding to each other, breathing, eating, running, flying, swimming, reproducing, nest building, etc. Among such actions and interactions of living things one finds plenty of evidence for vital phenomena, forces, laws and properties that are not to be found anywhere among inanimate objects nor among the molecules of which the living are constituted. In other words, the special vital forces that distinguish living things from the non-living are emergent, holistic properties of the living entities themselves. They are not properties of their physico-chemical components nor can they be fully explained merely in terms of physics and chemistry. This does not mean they are in any way supernatural or mystical. Those who conceived vital forces in supernatural terms were just as wrong as those who denied their existence. These higher, vital holistic phenomena and properties of living things are just as real, just as cause-effective and deserving of scientific recognition, as are the properties and laws of molecules or atoms, or electrons and protons.

When reductionist doctrine tried to tell us that there are no vital forces, just as it also had long taught that there are no mental forces, materialist science was simply wrong. Biological theory in this case was concentrating on the mass-energy of material components of living things and neglecting to appreciate the role of the non-material space-time components which also are critical. In anything living or nonliving, the spacing and timing of the material elements of which it is composed make all the difference in determining what a thing is.

The non-material space-time components, even when recognized, tend to be thrown out and lost in the reduction process, as science aims toward ever more elementary levels of explanation. If we think of things in terms of a mass-energy, space-time manifold, it can be seen that the space-time infrastructure gets short-changed in our traditional mass-energy interpretations.

The modern molecular biologist is quite willing to recognize the power of chemical or molecular forces and to grant scientific respectability to the laws that describe their interactions, even recognizing the critical role played by the inner spatial and temporal configurations. When the entities are no longer molecules, however, but living organisms, the reasoning suddenly undergoes a flip-flop change.

For many decades, science has been teaching that we and our world are composed of nothing but aggregates of electrons, protons and other subatomic elements. This overlooks the fact that it is the differential, non-material spacing and timing of these elements, as much as the material elements themselves, that mainly cause the world to be what it is.

COMMENTARY

Dr. Sperry was referring just now to the microscopic level of subatomic particles where the non-material spacing and timing of the particles makes a critical difference in determining exactly what a thing is. In nuclear physics and chemistry, for instance, internal space-and-time factors determine whether one is dealing with the ordinary element strontium, used in things like color-T.V. tubes and crimson fireworks, or the extremely hazardous strontium-90, a radioactive isotope present in the fallout from nuclear explosions.

But similar factors apply at the macroscopic level of social interaction as well. Though it may sound trivial, it is precisely those non-material elements of space and time at work in a complex system like freeway traffic that determine whether we have persons driving in a freely-flowing pattern or persons being driven by clogged lanes and jams. Congested traffic prohibits one's freedom to move, change lanes and travel at a self-regulated pace. Fluctuations of time and space have altered the "system" so much that a new system (the "traffic jam") has begun organizing itself, one that intensifies competition among the cars and substantially alters the reality of what exists on the freeway. The basic material elements of the two systems are the same: cars, drivers, asphalt and concrete. But fluctuations in the non-material elements of space and time have spawned something new (and, for weary workers returning home, something exasperating!).

This new reality affects not only the external system of freeway traffic, but the psychological interactions of the drivers as well. In a traffic jam other drivers often become aliens who threaten our turf or enemies who need to be vanquished. Changes thus appear throughout all the interlocking systems, and new rules governing those systems emerge as well. This basically supports the point that new forces and laws may emerge which cannot be explained simply in terms of elementary natural laws. Is that why, Dr. Sperry, you've suggested that the lower level forces in any entity are enveloped, overwhelmed and overpowered by the the higher?

DR. SPERRY

Yes, the point here is not only that new forces and new laws of the universe emerge at higher levels and that the higher cannot be fully explained or understood in terms of the lower, as has frequently been noted in the past—nor even that it is largely the new non-material space-time factors as well as the material components, that determine the nature of reality.

The further point that changed all this story in the past decade, from the status of occasional philosophy and minority science to that of the practicing dominant doctrine in psychology is the new stress on causation, i.e. the idea that in the reciprocal interaction of lower and higher levels the higher laws and forces (once evolved) exert downward causal control over the lower forces. The lower level forces in any entity are enveloped, overwhelmed and overpowered by the higher.

In scientific theory this means that the trajectories through space and time of most of the atoms on our planet are not determined primarily by atomic or subatomic laws and forces, as quantum physics would have it, but rather are determined by the laws and forces of chemistry, of biology, of geology, of meteorology, of psychology, even sociology, politics and the like. The molecules of all higher living things, for example, are not moved around in our biosphere so much by molecular laws and forces as they are by the living, vital powers of the particular species in which they are embedded. Such molecules are flown through the air, galloped across the plains, propelled through the water, etc., not by molecular forces (nor by quantum mechanics) but by the specific holistic vital properties possessed by the organisms in question.

Much of this seemed a matter of common sense and direct observation until science came along and began telling us otherwise. Ever since, there has been a growing conflict of worldview between scientists and the rest of society. The conflict is felt most acutely among the humanities and especially among those disciplines most concerned with moral values. What we are saying here seems to be, in effect, an admission that the humanities and common sense were on the right track all along in these matters while we in science were misled.

The errors are now being corrected, however, and any differences in language, ideas and beliefs that remain between scientists and the rest of society are not different in kind from those between two distant sciences. The profound conflict of worldview disappears.

COMMENTARY

You seem to be saying, Dr. Sperry, that just when science thought it had the structure and behavior of the universe "figured out" through the laws of atomic chemistry and quantum physics, scientists like yourself began to suspect a wrong turn had been made. What moves the molecules of higher living things is not so much molecular laws and forces as "the living, vital powers of the particular species in which they are embedded." Your description of such molecules "flown through the air, galloped across the plains, propelled through the water" evokes a passage from Walt Whitman's "Song of Myself,"

> A gigantic beauty of a stallion, fresh and responsive
>   to my caresses, ...
> His well-built limbs tremble with pleasure as we race
>   around and return. . . .

*What determines the relation between horse and rider is ulti-
mately neither their subatomic particles nor microscopic molecular
events. The molecules of complex living organisms like horse and
rider are moved by large macroscopic forces—mental intention,
speed, friction, weight, human response and animal instinct—to
say nothing of the skills and pleasures of equestrian sport. "New
forces and new laws" emerge at higher levels of life, and "the higher
cannot be fully explained or understood in terms of the lower."*

*But perhaps at this point some caution is to be recommended. To
speak of non-material vital powers or living forces that govern the
operation of our evolving universe may convey the impression that
the "new science" is in fact promoting a new species of mysticism
or religion. There is, indeed, some recent literature which speaks
rather mystically of an ultimate "cosmic essence"—charmed quarks
or dancing energy—linked with what traditional religions have
called "God." This literature relates the changes in worldview
brought about by new discoveries in physics, chemistry and biology
to theological speculation from both Eastern and Western sources.
Would you be willing to comment on this phenomenon, Dr. Sperry,
and to indicate how it differs from your own approach?*

DR. SPERRY

Certainly. The changes in science and worldview that
have resulted from research by me and my colleagues have
to be distinguished from the sometimes similarly described
—but actually quite different—renovations brought about
by recent developments in theoretical physics, referred to
in some places as the "new physics." In contrast to the
downward control concepts described here, the main theoretical change in physics adheres to the reductionist

approach (reducing to component parts) and is concentrated on the nature of the ultimate particles of matter as cosmic essence, suggesting that these ultimate entities are not so particulate, nor so separate as once thought, and are better described in probabilistic energy terms. These changed views of subatomic events have been very questionably extrapolated to the macroscopic realm as well by some writers, with analogies to Eastern religion and Taoism, inferring that macroscopic phenomena also are less material and machine-like than formerly supposed.

When physicists found that classical Newtonian laws didn't work any more for elementary particles but that a new theory, quantum mechanics, did work, they accordingly abandoned support for the old Newtonian doctrines in favor of the new quantum theory. The new theory was taken to be a more accurate and more comprehensive description of nature. I reject this on the ground that the subatomic properties, laws and forces, regardless of their nature, are superseded by forces operating at higher macroscopic levels. There is no way quantum mechanics could replace classical mechanics for things larger than molecules. Quantum theory cannot handle the pattern factor that the classical laws naturally incorporate. Neither is wrong; we need both. But for different things. If our thinking is correct here, it is not legitimate to extrapolate from the nature of subatomic events to the world at large. The emergent entities at higher levels contain, envelope and control the properties and expression of the elementary particles. So the common world at the macroscopic level is better described in the framework of biology, geology and the other sciences. The world is not all dancing energy or "charm" just because the ultimate building blocks seem to be of this nature.

Materialistic thinking commits similar errors, when in line with reductionist doctrine, it teaches that the forces and laws of the universe are blind, impersonal, purposeless and uncaring. Among all the forces that impinge on mankind affecting our welfare and future, none is of more prominent and critical importance than the forces of human society by which we are surrounded and which, of course, are often personal, caring and replete throughout with purpose. The kinds of forces embodied in society, in family, friends, politics, legislation, urban development and all the rest, including the expression of ethical, moral, and religious values, are all part of the natural order. Even below man, evolution as it progresses acquires a directionality and a complex self-built design with higher level controls that hardly fit the old mechanistic concept of a blind purposeless machine. Evolution can be viewed as a gradual emergence of increased purposefulness among the forces that move and govern living things.

COMMENTARY

Two points are suggested by what Dr. Sperry has said. The first is that the attempt to extrapolate theological conclusions from developments in theoretical physics is mistaken because, among other reasons, the new physics adheres to the reductionist approach by seeking "ultimate particles of matter" as the key to understanding all that goes on in our universe. Secondly, he suggests that the older materialist view in science falls from another direction into the same reductionist trap by teaching that the ultimate laws of the universe are uncaring and purposeless.

Against the view that evolution must result in mechanistic development or entropic disorder and loss of purpose, he proposes to define evolution as the simultaneous self-emergence of complex design and increased purposefulness among the realities that move and govern living things. This permits him to argue both that the universe is personal, caring and purposeful, and that the forces embodied in such things as social life, politics and religious values are all part of the "natural order." Such a position thus avoids both the reductionist trap and the dualistic supernaturalism which assigns such matters as ethical values, morals and religious belief to a strictly other-worldly source.

A number of implications flow from Dr. Sperry's view. Ours is a universe of endless innovations and creativity. And this is so not only for realities outside us, but also for our minds and consciousness. The mind constantly reorganizes itself, thrusts itself into new patterns that further enhance its potential for learning. Sometimes these shifts in the self-organizing mind may occur suddenly and dramatically, as when we are jolted into new awareness, flooded by new insight, or converted to an entirely fresh vision of the world and our role in it. Such changes are neither blind nor purposeless. They are in harmony with the larger evolutionary pattern of the universe itself, a pattern that is personal and replete with purpose.

DR. SPERRY

The point is that human nature and these higher kinds of controls in nature don't reduce any more to physical and chemical mechanisms, but have to be reckoned with now in their own form, in their own right. Vital, mental, social and

85

other higher forces, once evolved, become just as real as the evolved forces of molecules and atoms and must be given their due, over and above the elementary physical components. It will be evident that any theology that perceives God as equated with, or imminent in, the "laws and forces of the universe" comes out on these terms with a set of values and beliefs very different from those based in the traditional reductionist interpretations of materialist science.

The creative process in evolution involves control variables, forces and pressures operating at many different levels from the sub-molecular up to the ecologic, meteorologic and even astronomic in that the sunlight, seasons, phases of the moon, tides, etc. are all ultimately involved. The whole process depends on genetic mutations at the molecular level and, although the physical chemist might not agree, we can concede with the French biochemist Jacques Monod that the genetic mutations are a matter of chance at least from biological perspectives. But this does not mean, as Monod and other reductionists infer, that the whole process and course of evolution is governed ultimately by chance.

Most of the "chance" mutations prove lethal and are disposed of, not at random but according to the way they fit or don't fit into the developmental design of the species in question, itself a complex product of eons of evolution. Among the few mutations that survive the developmental constraints, there are many more "natural selection" pressures which control the further survival and fate of mutants that also are not matters of chance but products again of eons of acquired evolutionary design.

Among these higher selection pressures that include the competition for mates there are pressures at work that move the creative process toward ever improved, more competent, more attractive and more diverse life forms. Even beauty is selected for, as in mate preferences and in flower preference among pollinating birds and insects, etc. It is these higher laws and forces at the organismic, ecologic, and still higher levels that are in command in the creative process as much as or more than the events at the genetic level. It may all have started initially at the molecular level but as the process evolves, it incorporates space-time design, pattern and form factors at higher levels that, once established, become just as real and causal as those at the molecular level.

One can agree that the scientific evidence speaks against any preplanned purposive design of a supernatural intelligence. At the same time the evidence shows that the great bulk of the evolving web of creation is governed by a complex pattern of great intricacy with many mutually reinforcing directive, purposive constraints operating at higher levels, particularly. The "grand orderly design" is, in a sense, all the more remarkable for having been self-developed. To deprecate the higher emergent properties on the basis of their initial elemental building blocks is to further the error of materialistic thinking.

COMMENTARY
Both the supernaturalism of traditional theology and the reductionist strategy linked to the old scientific materialism and the new theoretical physics seem excluded as avenues for understanding

what or who constitutes the "grand orderly design" in our universe. Dr. Sperry has suggested that the intricate pattern which governs "the great bulk of the evolving web of creation" is a self-developed one which emerges from the very process of evolution itself. Mind and mental forces are surely a significant part of that self-developed pattern. But is it in any way legitimate to link mind, in this sense, with that "supreme intelligence" traditionally called God (or perhaps "Mind")? Would such a linkage merely signal a relapse into dualistic religious views which place the evolving universe in one sphere and God or Mind in an other-worldly "elsewhere"?

Or is there perhaps another way to deal with these questions? There is religious belief and mysticism. The publishing industry has been hit by a veritable explosion of works in this area. What has sparked this extraordinary growth of interest?

DR. JOSEPHSON

This explosive growth was started off by a book by Fritjof Capra, *The Tao of Physics,* to which Dr. Sperry has indirectly referred. Capra is a scientist working in the field of high-energy physics, who has himself gone through techniques to explore mystical experience. So his book, more than many others, is based on knowledge of both realms, and in this book he describes the many connections, or corresponding similar patterns, linking modern scientific discoveries and features of mysticism. Capra's book, it should be clearly stated at this point, is in no way (as has been suggested rather often by reviewers lacking knowledge of mysticism) an exercise in which somebody looked through what scientists said, looked through what mystics said, and thought ah! this quotation looks like that, so there's a connection. It certainly isn't that. Capra is very familiar with what is involved in both fields and I believe that the connections he wrote about in his book are deep and fundamental.

Capra's connections are of a very general kind. The scientist would like to know the connections at a more detailed level. Capra in fact was rather pessimistic about the possibility of making a detailed connection between science and mysticism: he says that mysticism deals with the roots of reality while science deals with the branches, and these are totally different things and we shouldn't expect to be able to connect them. My own belief is that mathematics may come to our aid and allow us to connect these two things. It is one of the striking features of science that mathematics is able to join together extremely diverse phenomena. Therefore I feel that Capra is being pessimistic in saying that we will never succeed in joining science and mysticism together.

Another book which has recently appeared, *Wholeness and the Implicate Order,* by David Bohm, embodies to some extent the initial stages of this process. Much of this book is qualitative, but Bohm and co-workers are trying to work on the mathematics involved, and I think that they are making progress. Bohm doesn't talk much about mysticism explicitly in his book, but nevertheless it is very much based on mysticism. He approaches things from the direction of science. For many years he has been concerned with an attempt to resolve some of the paradoxes thrown up by quantum mechanics, some of which I shall be describing. These present great difficulties of a philosophical nature.

Most scientists just sweep these philosophical difficulties under the rug, but Bohm wasn't satisfied, and he thought hard about the implications of these paradoxical aspects of quantum mechanics.

Let me say a few words about these paradoxes. One of them is that in quantum mechanics we seem not to be able to describe what goes on in a fully deterministic way; in other words we seem not to be able to explain what goes on in mechanistic detail. The second kind of paradox is that separate regions of space seem to be connected together in ways which one cannot readily understand in an ordinary kind of mental picture. Bohm, trying to understand these things, came up initially with the idea of hidden variables. What he proposed is that the fact that things don't seem to be well determined in science means that there are variables which we can't observe directly, which do however affect the physics. His views gradually evolved, and he says now that what these paradoxes show is the existence of an unmanifest or implicate order which we can't observe directly. Phenomena are created from this order rather in the same sort of way in which a cloud condenses from moist air. We see order in the observed phenomena which is the result of the unobserved order.

Thus Bohm views quantum mechanics as providing an indication that there is unobserved order in Nature.

COMMENTARY

This suggests a connection, doesn't it, between what Bohm describes as an "implicate order" and the notion of God as an intelligence who works "behind the scenes," as it were, to bring order from disorder?

DR. JOSEPHSON

If we want to put God or Mind into science then we have to say that there is an intelligence behind the scenes, which is creating order, or at least leaving things less disordered than they would have been without the intelligence being present. And so we can identify the unobserved order with intelligence. Here we are on the way to the beginnings of a mathematical synthesis. If what Bohm is doing with un-manifest order can be combined with the mathematics of intelligence, we'll be well on the way to integrating intelligence into the framework of science.

COMMENTARY

One gets the impression, Dr. Josephson, that your hypothesis concerning a possible combination of Bohm's implicate order with the mathematics of intelligence would lead to an inclusion of God or Mind within the framework of science in a manner which exceeds what Dr. Sperry is willing to grant. Sperry has invited us to understand the order and purpose that permeate our evolving universe as self-emergent consequences of the process of evolution itself. As he noted earlier in this Conversation, scientific evidence "speaks against any preplanned purposive design of a supernatu-ral intelligence;" indeed, the "grand orderly design" we perceive in our world is "all the more remarkable for having been self-developed." Sperry's view differs rather considerably, it seems, from your conviction that if we include God or Mind in science, it would make a difference in the way we understand order emerging in the universe.

DR. JOSEPHSON

I'm going to be venturing farther than Dr. Sperry did and deal with some controversial issues. Paradigm shifts, such as may be involved in the current issue, very often involve a good deal of controversy (as for example, when Einstein introduced relativity, and a number of ideas which people thought were absolutely correct and unchangeable had to be given up).

Now before I get down to any actual detail, I'd like to consider some background questions such as what the relationship of science to religion should be anyway. As was mentioned by Dr. Sperry, a couple of centuries ago science and religion started to diverge—each wanted to take over the whole territory of knowledge to itself, science thinking everything could be explained scientifically, and religion, possibly as a kind of reaction to what the scientists were doing, believing everything could be explained in religious terms. And so we have a situation at the present time where there is essentially zero overlap between science and mental phenomena: you don't find in a scientific paper somebody saying (for example), such and such a phenomenon was an instance of your "mind enfolding matter." So just from the fact that scientific work makes no mention of God or Mind we see that science and the mentalist revolution are at the present time totally separated from each other. But, as you have already said, things are beginning to change and overlapping views are beginning to be found. However, the fact remains that science gets on quite well without God, and perhaps we should look into the reasons why this might be. If we assume God or Mind does exist, then why hasn't this appeared in scientific experiments? I'd like to indicate two factors which may be relevant.

Firstly, science casts the spotlight which it uses to search for knowledge very selectively; in other words what scientists choose to look at, to try to explain in scientific terms, is rather restricted, rather biased. And the content of science is biased in a materialistic direction. This applies to almost all the sciences, the physical sciences as well as the biological sciences. The reason is very largely that it is easier to study quantitatively the behavior of matter and the grosser aspects of behavior (both animal and human), than it is to study higher human behavior where the influence of God might be significant. So science, in choosing the simpler problems to examine first, tends always to look in directions where theological concepts are not very relevant.

Secondly, even within a particular field science likes to look at simple phenomena, as these are more easily connected with fundamental laws. Then one tends to say "We can explain the simple phenomena very well now; eventually, we'll be able to explain the complex phenomena as well." The gap between simple and complex phenomena is one which scientists tend, just as a matter of faith, to assume (especially if they are of materialistic orientation) will be bridged without invoking any higher being. For example, the question of how man came into existence is assumed to be a problem which will be solved in the future when we have filled in all the details (and again, it is supposed by people working in artificial intelligence that problems of higher human skills will be solved in just a matter of twenty years or so more research). But perhaps this is questionable, perhaps this gap cannot be bridged. A recent book by Richard Thompson, *Mechanistic and Non-mechanistic Science,* goes into the difficulties which conventional science has in explaining phenomena like evolution and creativity.

I've been talking so far about the materialistic orientation of science. Now, there are two ways in which one could approach the issue of whether God has an influence on Nature. One is to continue following the traditional materialistic line of explanation, seeing if it really does explain everything. That would be a very long job—it might be a couple of centuries before we would get an answer that way. An alternative approach for the scientist is to say, Let's investigate the opposite view, i.e., that perhaps we should be taking God or Mind into account in science; what would a science look like which had God in there playing a part, accounting thereby for particular phenomena?

There are various ways into this problem, and the way I'm going to take is to say that if we want to put God or Mind into science, then the primary feature of Mind, the one which is most closely connected with the science we've got, is intelligence. What I want to suggest is that the new science will start by understanding and describing "being intelligent." One of the reasons for thinking this to be a good thing to do is that intelligence as a phenomenon in itself is inadequately dealt with in most branches of science except perhaps cybernetics. Intelligence is something which in science is generally studied in terms of details, and not as a general phenomenon. There is a kind of gap here in our world-view; filling it in may give an expanded world-view, and into this world-view the supreme intelligence may fit naturally.

COMMENTARY

Dr. Josephson has proposed that the inclusion of God or Mind in science is not only plausible, but may even be necessary if science is ever fully to understand Nature or to overcome its difficulties in explaining phenomena like evolution and creativity. Such a new approach in science will start, he suggests, by understanding and describing "being intelligent," since there seems to be a correlation between the unobserved "implicate order" in Nature and the mathematics of intelligence.

DR. SPERRY

We should rather start by first considering a more general, preliminary question, one that usually raises the greatest popular concern, namely, what would remain on which to build the mentalist revolution if we were to fully accept the world view of science and therefore to exclude everything that science disavows? This would seem to require firstly an exclusion of all supernatural and otherworldly forms of existence for which the empirical evidence and scientific progress seem increasingly to disclaim. In other words, if we eliminate ghosts and angels and otherworldly forms of deity, devils and dryads and spirits of all kinds, myths of heaven, hell, astrology and the hereafter, witchery, the occult, the mystic, the paranormal and everything else that modern science rejects, what would we have left to believe in?

The answer, of course, is: plenty—especially on our revised mentalist terms. No one yet has described another realm of existence, creation or creative forces that even remotely compares in the vastness, complexity, diversity, wonder, and yes, beauty and meaning, with the real world revealed and described by modern science (including the human and social sciences). On our current revised terms that emphasize self-emergent holistic and transcendent qualities, the insights of science give added, not lessened, reasons for awe, respect, and reverence.

COMMENTARY

Dr. Sperry seems to be saying that the exclusion of "all supernatural and otherworldly forms of existence" would destroy neither the "downward" causal control of mind over brain functions, so important to the mentalist revolution, nor the transcendental and personal qualities, the purpose, complexity and beauty evident as self-emergent phenomena in our evolving universe. This would appear to render "God" or "Mind" an idle speculation which does not advance the investigations of science. Who needs to invoke God, who in the traditional view presides over reality from outside space and time, when the creative forces of the universe itself are sufficient to invest life with beauty and meaning, mystery and reverence?

DR. JOSEPHSON

Let me now put my view on a more concrete basis, by going over an old theological line of argument with which you are probably familiar. This is the argument that one can see that God exists through the design in nature—everything in nature functions in a very precise manner, just as if someone had set it all up by writing blueprints and saying this is how the world should be. This is the old theological viewpoint, that one just sees, looking round one, that the only reasonable explanation for all this is that there was a designer who made things this way. Now of course this viewpoint is the one which has been under attack by science for a long time, because science has attempted to show that everything that had been explained by religious people by invoking God could be explained in scientific terms instead; for example the existence of man is "explained" in terms of the evolution of species.

Now, the question is, does this kind of explanation actually explain facts like the existence of man? Consider the following situation: a house is being built, and a scientist on Mars is looking through a powerful telescope at what is going on (the bricks moving about and being assembled into the structure of the house, etc.). He might well say "I can explain all this. People's muscles are contracting because of impulses coming down to them from the brain, this makes the arms move about and as a result the bricks get lifted. I can go through my calculations and see that the motion of every brick is exactly as given by the laws of mechanics, the laws governing the transmission of nerve impulses and the contraction of muscles and so on." So science has explained everything; it has explained the building of the house, everything just follows the ordinary laws of physics and we don't have to bring in the idea of intelligence at all.

Now, we know in this case that the explanation is incomplete, or at any rate misleading: it is the human being's intelligence and his knowledge of how to move things that are responsible for the house being built and a precondition for the whole process to be possible at all. In addition, the arrangement of bricks is a consequence of the existence of a blueprint which somebody made because he knew how to design houses, and so on.

It is quite possible that the current scientific explanation for the existence of man may be equally inadequate. When intelligence is present we don't decide on its presence or absence just by seeing whether the laws of physics are obeyed; intelligence is not like a new energy source. The presence of an intelligence manifests itself via the presence of or the creation of states which are a priori extremely unlikely: states such as all the bricks fitting neatly to form a house, all being put together in the right way. That is, intelligence manifests itself by making certain unlikely situations appear. And this is the sort of thing that would be studied in a very general theory of intelligence. In other words, if we develop the relevant mathematics, as I suggested earlier, we might be able to see certain general principles at work, and we might then be able to see that the fact that such and such happens suggests the presence of an intelligence—it would be too unlikely to have happened by chance. So, the way this approach would go through, if it were possible to carry it through, would be to say that the principles of intelligence are universal.

These explorations would be different in one interesting respect from ordinary scientific explorations. Much of the research would involve not finding new knowledge, but verifying what mystics have already said. This is because mysticism is very highly developed already: sources like the Vedas and the Kabbalah have said a great deal about the nature of the regions to be explored, and so a scientist trying to explore these regions will largely be covering known territory. However, he will be covering it from a different point of view and perhaps trying to describe it mathematically.

◄§   §►

SUMMING UP CONVERSATION TWO

We've arrived at the end of the second Conversation held at the Isthmus Institute in Dallas. And once again we are confronted by an unresolved question: does the notion of "God" or "Mind" belong in any way to the framework of science? If we are adequately to understand the relation between mind (as interpreted in Conversation One on the mentalist revolution) and nature ("creation," the evolving universe, the entire cosmos of "matter"), is it necessary or desirable to appeal to the role of an order-creating, universal Mind?

In fairness to the positions of both Dr. Roger Sperry and Dr. Brian Josephson, it should be pointed out that neither scientist in this Conversation is attempting to address the specifically theological question "Does God exist?". The question they grapple with is: what are the best means available to describe and interpret the data that Nature and

the evolving universe, in which we are conscious participants, present to our senses and to our intelligence? The
question so posed is one of science rather than one of theology, though this does not exclude the possibility of convergence between these two modes of human inquiry. To bor-
row a phrase from the late American writer Flannery
O'Connor, Sperry and Josephson are attempting "to render
the highest possible justice to the visible universe."

The difference of position between Dr. Sperry and Dr.
Josephson should thus be understood not as a philosophical
conflict between atheism and theism, but as a difference of
opinion over the means that are actually available to science
in its effort to interpret nature, creation and evolution.
Josephson is willing to include among those means a "God"
or "Mind" who supremely exemplifies the kind of intelligence which reveals itself by making certain unlikely situations appear. If such an Intelligence is omitted from science's account of nature, Josephson fears, the story will be
incomplete and, indeed, misleading. Sperry, in contrast,
argues that we should rigorously adhere to the demands of
a science which disavows otherworldly spirits and deities as
explanatory factors in the evolving universe. In so dispens-
ing with "God" as a means of scientific interpretation, we
lose none of the vastness, wonder and complexity of existence.

Perhaps an approach to resolving some of the differences
between Sperry and Josephson may be found by turning
our attention to "Time," the subject of the "Interlude"
which follows.

I live my life in growing orbits
which move out over the things of the world.
Perhaps I can never achieve the last,
but that will be my attempt.

I am circling around God, around the ancient tower,
and I have been circling for a thousand years,
and I still don't know if I am a falcon, or a storm,
or a great song.

From "A Book for the Hours of Prayer,"
by Rainer Maria Rilke;
translated by Robert Bly

INTERLUDE I

৪৶

TO THINK OF TIME

৪৶

GUESTS AT THE ISTHMUS INSTITUTE

COMMENTARY

Interlude One is an editorial condensation and extrapolation of comments made by guests at the Isthmus Institute in Dallas after discussions that followed Conversation One and Conversation Two.

৬৪   ৪৶

During the years 1899–1900, the young poet Rainer Maria Rilke made two extended journeys to Russia. Rilke's poem "I live my life in growing orbits," which introduces this "Interlude," was harvested, along with others in the collection "A Book for the Hours of Prayer," from his experiences of Russian life and culture during his months as a tourist there. Rilke went to Russia, however, not only as a tourist but as a pilgrim. Described by one of his recent biographers as "a melancholy atheist with a guilty conscience," Rilke was fascinated by the "Russian" God, a deity whom he regarded as still evolving, waiting to be created by peasants and children, especially through nature's objects and art's icons. In contrast to the God of western Christianity, usually depicted as timeless, serene and utterly beyond nature, the Russian God of Rilke's imagination was passionately engaged in the time-bound world of natural life and ordinary people. Though he was no mystic in the ordinary western sense, Rilke was deeply convinced that somehow human persons and God were connected by an interdependent evolution—and that time, "the hours of prayer" punctuating days and seasons, was a key to understanding that evolving relationship. God, the "ancient tower" of Rilke's poem, is still a-building, as is the circling "I," whose identity and nature remain ambivalent and undefined. Only time unfolding will reveal whether "I" am falcon, storm or great song.

If time is an enfolding reality which discloses the co-dependent evolution of both humans and God (or "Mind"), perhaps it may also provide a key for understanding some of the convergences and some of the unresolved questions that have emerged from the first and second Conversations at the 1982 Isthmus Institute in Dallas. The points of convergence may be summarized first. We have heard Nobel Laureates Dr. Roger Sperry and Sir John Eccles agree that mind as well as brain is an irreducible reality. Yet the physics and chemistry of the brain cannot, taken by themselves, account for the complex phenomenon of mind, with its thickly populated world of ideas, ideals and intentions. As Dr. Sperry has observed, mind may be understood as self-emerging from the activity of the living brain, but the world of mental phenomena, intelligence and sensation, though functionally dependent on quantum physics and molecular chemistry, cannot be reduced to them. In short, mind has a real existence of its own, and we are compelled by the current scientific evidence to say that it exerts "causal control" over brain activities (Sperry) and that non-material forces thus act upon material ones (Eccles).

The new science proposed by these Nobel Prize winners thereby excludes both materialism of the sort which denies the reality and influence of non-material forces in the universe and reductionism of the technical kind that attempts to explain the activities of all living organisms by the same laws chemists and physicists use to interpret inanimate matter. But the new science does not merely exclude positions of the old. On the positive side of the ledger, Dr. Sperry and others are advancing a revised concept of evolution as "a gradual emergence of increased purposefulness among the forces that move and govern living things." This represents a radical departure from the older view which virtually equated evolution with entropy, the inexorable winding down and loss of energy that are supposed to accompany the aging process of the universe. Far from winding down, the universe, as interpreted by the new science, is ever expanding and irreversibly becoming; it never goes backwards. The world becomes—and our own experience can confirm this point—increasingly complex and beautiful, ever more full of meaning and purpose.

At these increasingly complex and purposeful levels of life, vital forces, as Dr. Sperry has referred to them, are working. These forces defy reduction to the level of atomic or subatomic laws and forces. This is because, as Sperry has noted, "the emergent entities at higher levels contain, envelope and control the properties and expression of the elementary particles." As the process of evolution unfolds irreversibly through time, the result is an enlargement of vitality and an increase of purposeful, self-organizing complexity among living things.

Further, the energy which is created and released by this increasing vitality builds, "overloads," and must consequently erupt into newer, more complex patterns of life. The universe we inhabit as conscious participants and observers is thus characterized less by entropy than by a vital becoming. Order emerges from disorder, purposefulness comes from random chaos.

From the moment of the Big Bang until now, the randomness, fluctuation and chaotic disorder of the universe have been evolving through irreversible time into self-organizing pattern, order, complexity. For ultimately, the most astonishingly lively of the "vital forces" Dr. Sperry has drawn attention to are vitalistic "energy" and "time." The interactions of these two primordial forces—time and energy—result in self-emergent patterns. Some of these patterns we call "stars;" others, "rocks;" still others, "people."

Although chemists and biologists may want to divide these patterns into more precise subdivisions of "animate" vs. "inanimate" or "organic" vs. "inorganic," the ultimate source of all self-organizing systems in our universe, from simple cell to complex brain, is the same: vitalistic energy interacting with time. Considered in terms of this source, divisions of such systems into living and non-living blur and lose significance.

Suppose you try drawing a rumpled mass of ribbons through a spool in which you have drilled a small hole. As the energy of pulling is exerted, and as the ribbons move forward through time, they will reorganize themselves. The ribbons may flatten out into a single taut string; they may split into separate strands, some of which curl into new patterns; or if the tension of pulling is relaxed, the ribbons may fall into lazily looping spirals. Through the interactions of time and energy, the ribbons organize themselves into a new pattern quite different from the rumpled mass on the other side of the spool.

Or, in a far more complex example, consider the light and radiation that continue to reach our earth from a star that has become extinct. The original configuration of mass, time and energy from which that light and energy emanated is no longer a "star" in the standard sense of that word. Evolving forces of time and energy have caused the once-proud star to become a white dwarf or a black hole. Yet the light and energy released by this process of stellar evolution do not simply get lost; they continue on the irreversible evolutionary course. New patterns and configurations emerge, as when some of that light participates in a process of photosynthesis to produce new shoots on a green growing plant.

Time is thus a primary key to understanding the irreversible process of evolution in our universe, to grasping how order emerges from disorder, complexity and purpose from chaos. Contrary to the view long common in the West, that time is a quantified object easily divisible into discrete parts of past, present and future, the new science advanced by the Nobel Laureates represented in these Conversations challenges us to experience time as an irreversible process of becoming. Time is not object but "event;" it is creative becoming rather than static being. Or, in the words of French poet and critic Paul Valéry, time is "construction."

This new view of time suggests that it is neither an artificial construct devised to impose an intelligible (but actually unreal) order on a universe in which chance and chaotic disorder reign, nor an inert and passive container within which events happen and natural objects appear. Time, as interpreted in these Conversations, is a creative, ever-becoming quality, the "condition of possibility" without which there could be neither an evolving universe nor evolving life. It is thus an intrinsic quality enfolding all that has become and will become.

Time has acquired, therefore, a new and perhaps unexpected status among scientists. Though he did not dwell on it explicitly, Dr. Roger Sperry has already called our attention to the importance of time. One of his criticisms of reductionist procedure in science, as he remarked in Conversation Two, is its tendency to neglect non-material space-time components when trying to account for the structure and behavior of reality. The very nature of reality, he said, is determined not merely by atomic and molecular materials, but by space-time factors as well. What and how a thing "is" results not merely from its constituent material elements—so many atoms of oxygen, so many atoms of hydrogen—but from the differential and non-material spacing and timing of those elements.

The new status accorded time has also caused scientists to reconsider their interpretation of such basic elements as "matter." Recent scientific research has overturned the popular view that matter—the subatomic, atomic and molecular "building-blocks" of the universe—is a fundamentally passive substance, subject to very strict, unchanging laws of chemistry and quantum physics. The world of microscopic reality seems now not to be so neat and orderly. That world is far more spontaneous—not to say capricious—than science at first suspected. Realities at the atomic and molecular levels are caught behaving in ways that appear to transgress the supposedly iron-clad laws of quantum mechanics. Indeed, atoms and molecules display properties formerly associated only with large, complex biological systems. Molecules appear to have a mind of their own; they seem to "feel," to remember and communicate with each other over distances far greater than the infinitesimal "spaces" that separate them. In the famous Belousov-Zhabotinskii reaction, for example, a homogeneous chemical solution undergoes radical changes of patterns fluctuating between disorder and order, yet the molecules of this solution "remember" the sequence in which those patterns occur and become highly sensitive to the tiny effects that lead to the selection of those patterns.

In other words, in terms of time and energy, the same basic vital forces which operate at the macroscopic level of complex organisms seem to be just as important for understanding what goes on at the level of subatomic particles, atoms and molecules. The irreversible arrow of time controls not only the self-organizing evolution of large organisms and complex systems like the brain and the human mind, but self-emergent becoming at the atomic and molecular levels as well. At all levels of the universe we see irreversibility, the creative "arrow of time," emerging out of disorder, randomness and instability.

Science has thus begun to see that time makes a critical difference in "what is" and in "what is real," whether one is dealing with the nature and behavior of molecules and atoms or the nature and behavior of large organisms. That is, if an irreversible arrow of time exists on the macroscopic scale—as our everyday experience of aging bodies and growing cities reveals—then the same kind of thing is going on in the microscopic world. As Dr. Roger Sperry's theory of "downward causation" implies, the laws and forces at work on the macroscopic scale supersede, envelope and enfold lower-level activities on the microscopic one.

This is an extremely vital and revolutionary point of view. In the past, science has been accustomed to thinking "from the bottom up." The newer science suggests, however, that we need to interpret nature and reality "from the top down." To put it more concretely, science was formerly in the habit of saying that what happens on the macroscopic scale of complex organisms is caused by what happens on the microscopic scale of subatomic and atomic events. This led scientists to regard mind as merely an effect produced by electro-chemical activity in the brain, with the material brain, as it were, "holding all the cards" and giving all the orders to the population of forces that occupy the human cranium. Without denying that mind self-emerges from the activity of living brain, we now have to admit that causation happens in the other direction, "downwards" instead of upwards. Once evolved, the mind exerts—to use Dr. Sperry's phrase—"downward control potency" over the brain's activities on the subatomic, atomic and molecular scales. At this point in our *human* evolution, in other words, it is more accurate to say that mind "causes" brain than that brain causes mind. And because the evolutionary arrow of time is irreversible, we cannot return to that pre-human "moment" which preceded the emergence of mind and of mind's enfolding action.

The implications of this revolutionary viewpoint are vast and pervasive. The evolutionary arrow of time has made mind not only possible but actual. Time has also made that mind conscious of itself, in the human species, and capable of reaching out to enfold lower-level activities within our evolving universe. Time has revolutionized the relationship between mind and brain, thereby revolutionizing the relationship between humankind and nature.

Until quite recently, the prevailing scientific model assumed that if we want to understand what's going on in human mind and mental activity, we must begin at the "bottom," with a microscopic analysis of chemical reactions and neural events, since these latter constitute and determine whatever may be said of "mind." The new science represented in these Isthmus Institute Conversations proposes a daring departure from this model by arguing from the "top." If we want to understand what's really happening on the microscopic scale of subatomic, atomic and molecular activity, we must begin with the enfolding potency of mind, which supersedes, but does not eliminate, those lower-level laws of microscopic activities. The macroscopic does not imitate or mirror the microscopic; rather, the reverse is the case.

It is thus not a perversely anthropomorphic arrogance that leads this newer science to propose that molecules feel, signal each other, remember and communicate over times and distances formerly thought possible only in the case of larger, more complex biological systems. It is precisely the enfolding phenomenon of mind and mental activity that sheds light on molecular structure and behavior. Somewhat paradoxically perhaps, non-material and purpose-laden mental activity, often described (and dismissed) as "the spiritual dimension" of human existence, disclose what actually governs most of the material realities in our world. With good reason Roger Sperry could say, in Conversation Two, that "the trajectories through space and time of most of the atoms on our planet are not determined primarily by atomic or subatomic laws and forces, as quantum physics would have it, but rather are determined by the laws and forces of chemistry, of biology, of geology, of meteorology, of psychology, even sociology, politics and the like."

The relation between "observer" and "observed" in the investigation of nature has thus been transformed. The very experience of ourselves as possessed of minds capable of sequenced thinking and sudden insight, blissful fancy and baleful foreboding, dogged determination and double-thinking deception, opens us to the possibility that the fundamental nature of reality itself is similarly quirky, unexpected, changeable. The illusory quest for a set of immutable laws which govern all aspects of the universe in a uniform manner must be abandoned. Ours is not a homogeneous universe. Even the pervasive force of time operates on a plurality of scales. In a scientific experiment, the human observer's time scale may differ vastly from that of the objects under investigation—as when a geologist examines rock formations that have developed according to the slow, patient time of geological evolution.

So we find ourselves, at one and the same time, as possessed of minds that exert "causal control potency" over lower-level realities *and* as conscious participants in the very universe we seek to investigate. We observers observe ourselves observing. But as "subjective" as this may seem, there remain constraints that identify us as part of the world we're describing—the speed of light in a vacuum, for instance, which limits the velocity at which any observer can transmit signals. Our dialogue with nature can take place only within nature.

Time and mind are, then, the two basic keys we need for understanding our evolving universe and our human future in it. The irreversible arrow of time evolving makes mind possible—and mind, once evolved, enfolds the lower-level forces and laws from which it emerged in the first place. The new science which takes all this into account begins at the "top," where mental forces supersede and envelope activities on the atomic or molecular scale. This blends in, too, with Dr. Brian Josephson's suggestion, in Conversation Two, that perhaps science should consider beginning at the "very top" by entertaining the possibility of a God or Mind which can be taken as an implicate order in the evolving universe. This order, though it remains unobserved, manifests its intelligent source through the presence or creation of states which are, a priori, extremely unlikely.

We thus return to the place where we began, to the evolving mind of that anonymous "I" in Rainer Maria Rilke's poem, as it circles through time around the "ancient tower" called "God." "I . . . move out over the things of the world," conscious that my relation to them and theirs to me is ever changing, ever in process of creation and renewal. Perhaps Rilke, the atheist with a wounded conscience, may serve as a kind of patron saint for the new science which starts at the "top" with the enfolding power of mind and seeks clues to an implicate order within a universe wrapped in time.

It is hard to avoid the impression
that the distinction between what exists in time,
what is irreversible,
and . . . what is outside of time,
what is eternal,
is at the origin of human symbolic activity.
Perhaps this is especially so in artistic activity.
Indeed, one aspect of the transformation
of a natural object, a stone,
to an object of art
is closely related to our impact on matter.
Artistic activity breaks
the temporal symmetry of the object.
It leaves a mark that translates our temporal dissymmetry
into the temporal dissymmetry of the object.

From *Order Out of Chaos,*
by Ilya Prigogine
and Isabelle Stengers

INTERLUDE II

࿇

THE REDISCOVERY OF TIME

࿇

BY ILYA PRIGOGINE

ABOUT INTERLUDE II

A certain convergence of view between the art of poetry and the art of science has already surfaced a number
of times in these Nobel Prize Conversations held at the
1982 Isthmus Institute in Dallas. In the following piece,
Dr. Ilya Prigogine, often called "the poet of thermodynamics," addresses the challenge presented to modern science by "the rediscovery of time." He does so not only as
a world-renowned research chemist whose work has been
awarded the Nobel Prize, but as a thinker sensitive to the
multiple interactions between the world of natural objects
and the symbolic world of the artist. These worlds inter-
sect in time, though the artist often struggles to register

through symbols the relations between what is time-bound and what is eternal. As Dr. Prigogine notes, in the passage from *Order Out of Chaos* that introduces this "Interlude," artistic activity "breaks the temporal symmetry" of objects, and "translates our temporal dissymmetry into the temporal dissymmetry of the object."

When we see objects in our everyday experience, we attribute the same "time" to all of them—all exist in what we like to call, for want of a better word, our "present time." So the rocks jutting from a ledge on the bluffs above the Mississippi at Dubuque are thought of as existing in a "time" perfectly symmetrical with the "time" of the corn stalks growing on farms just outside the city. Only rarely do we consider that the rocks on the bluff have an internal geologic time that stretches back many millions of years. The rocks exist in more than one time-frame. Nor are those frames symmetrical, either with each other or with the relatively brief life span of the corn plant. We begin to recognize that these apparently stable objects—rocks and growing corn—are caught up in an evolutionary arrow of time, just as we observers live in the irreversibly forward flow of time. Just as we do not experience time as a perfectly symmetrical reality, so we recognize and translate that experience into the life of rocks, corn plants, and the rest of those objects that inhabit our daily world.

Artists are especially sensitive to these fluctuations, these dissymmetrical times that enfold both us and everything we perceive. Art breaks the illusion of temporal symmetry—it happens most obviously, perhaps, in music—and confronts us with the enfolding power of time as it recapitulates past and anticipates future. Nature, that ensemble of beings, sounds, smells, sights and objects of which we humans are statistically so small a part, has the ability, through the interactions of time and energy, always to become fresh and new, always to begin. Artists perceive this fresh beginning by breaking the illusion of static symmetry. They are aware, as few of us ever are, that we live always in the first moments of a newly-becoming creation, not at its dwindling end.

Still, the exact nature of time has troubled great minds, both scientific and artistic, from Aristotle and Sophocles to Einstein and Stravinsky. But though all of us experience time daily as welcome potential, unwelcome intrusion, or some combination of the two, scientists remain divided in their interpretations of it. One thus cannot expect that the views of time presented by the different Nobel Laureates at the Isthmus Institute will all agree. Not only is the universe we share with other beings pluralistic and non-homogeneous, but scientific interpretations of it vary as well. We are still far away from any unified concept of evolution, the universe, the human future or the human mind.

Dr. Prigogine's analysis of time, originally composed for the Isthmus Institute and later presented to the American Academy of Religion in December, 1983, is included here without commentary.

ولاء   هو

ILYA PRIGOGINE

In the preface to the 1959 edition of *The Logic of Scientific Discovery,* Sir Karl Popper states that:

. . . there is at least one philosophic problem in which all thinking men are interested. It is the problem of cosmology: the problem of understanding the world—including ourselves, and our knowledge, as part of the world.

It is obvious that the meaning of time plays an important role in the problem so beautifully spelled out by Sir Karl Popper. It is therefore important to stress the fact that our vision of nature is at present undergoing a radical change toward the multiple, the temporal and the complex.

Till recently, a mechanistic world view dominated western science, a view according to which the world appeared as a vast automaton. We now understand that we live in a pluralistic world, whose description involves elements not included in the traditional picture.

It is true that there are phenomena that appear to us as deterministic and reversible, such as the motion of a frictionless pendulum, or the motion of the earth around the sun: reversible processes do not know any privileged direction of time. But there are also irreversible processes that involve an "arrow of time". If you bring together two liquids such as water and alcohol, they tend to mix in the forward direction of time, that is, in our future. We never observe the reverse process, the spontaneous separation of the mixture into pure water and pure alcohol. Mixing is therefore an irreversible process. All of chemistry also involves such irreversible processes.

Today we are becoming more and more conscious of the fact that on all levels, from elementary particles to cosmology, randomness and irreversibility play an ever-increasing role. Science is rediscovering time. This obviously introduces a new dimension into the old problem of the two cultures, science and the humanities.

Most of European modern philosophy, from Kant to Whitehead, appears as an attempt to overcome in one way or another the necessity of a tragic choice between the mechanical view of classical physics, and our daily experience of the irreversible and creative dimension of life. On this perspective I could only confirm the views expressed by Ivor Leclerc:

> In our century we are suffering the consequences
> of the separation of science and philosophy which
> followed upon the triumph of Western physics in
> the eighteenth century.

However, I believe that the situation today is much more favorable in the sense that the recent rediscovery of time leads to a new perspective. Now the dialogue between hard sciences on one side, human sciences and philosophy on the other, may become again fruitful as it was during the classic period of Greece or during the 17th century of Newton and Leibniz.

To illustrate this coming together on a fundamental point, let us consider in this lecture the relation between *Being and Time,* to take up the title of the influential essay of Martin Heidegger.

This relation may probably be considered as one of the central themes of Western philosophy. The aim of my lecture is precisely to point out that today we can envisage a fresh approach. Obviously, this relation does affect large parts of epistemology, and even ontology. I do not feel prepared to discuss the theological context; however, I believe that such a discussion will always encompass the new concepts science affords us about man's position in nature, and is therefore unavoidably related to a discussion of the problem of Being and Time, or Being and Becoming.

Let us start with a brief summary of the way in which time was described in classical Physics. Western scientific tradition takes for granted since Aristotle that Time is closely related to motion, and therefore to space. As a consequence of this view, we have inherited the idea of an isomorphism between time and a one-dimensional space, as shown in the classical representation of time, in which the present separates the past and the future.

This description is used in classical physics, as well as, with minor modifications, in quantum theory and in relativity. While immensely useful, it does not do justice to the various connotations of time. Past seems to disappear in the present. Present disappears in the future. No intrinsic connection appears between past, present and future.

Both our conscious experience and the existence of an evolutionary, time-irreversible universe seem to point to a far richer and subtler concept of time. We may imagine that at present we are sitting on a hill; how does it happen that we glide down always in the same direction? Why do we age all together?

We have therefore to reconsider the meaning of time. This, as is well known, was the conclusion reached by

Bergson, Whitehead, Husserl and Heidegger, to quote only some of the deepest thinkers of our days. However, in contrast with their approach, I want to show here that a new time concept can be generated from within modern science, and does not imply a complete break with the scientific tradition of the West.

Already Aristotle associated time with generation and corruption—in our modern language, to qualitative change not reducible to local motion. But it was only recently that this aspect of time could be expressed in a precise mathematical form. Let us start with the question about irreversibility, most closely connected with the problem of evolution.

The difficulties in the understanding of irreversibility show up very clearly in the classical approach of Boltzmann. Let us consider the entropy S, which is the basic quantity which, according to the second law of thermodynamics, increases in isolated systems as the result of irreversible processes. Boltzmann's great idea was to express S in terms of a probability P. This is the content of his celebrated formula, $S = k \log P$.

Here, k is a universal constant, the so-called Boltzmann's constant. As follows from this formula, in isolated systems, entropy S increases because the probability increases. At thermodynamic equilibrium, complete disorder is reached, and the probability is maximum. Boltzmann's formula is certainly one of the most important of theoretical physics. I have no intention of going into the controversies to which it has led, but I still would like to emphasize a basic conceptual difficulty imbedded in Boltzmann's attempt.

In modern probability theories a fundamental role is played by the so-called transition probability to go, at time t, from one point, say ω_0, to a region E in phase space.

(See the illustration, p. 125)

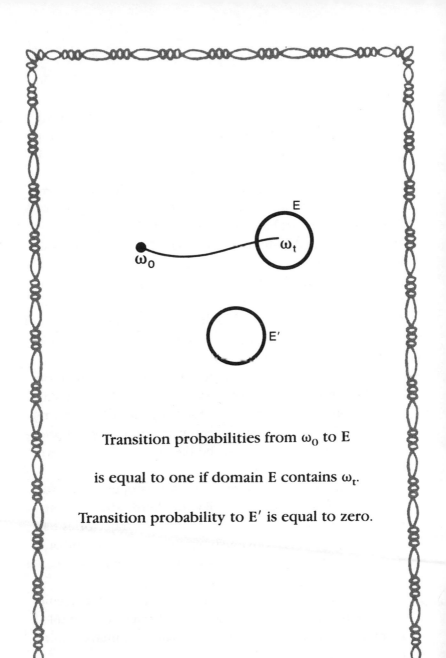

Transition probabilities from ω_0 to E

is equal to one if domain E contains ω_t.

Transition probability to E′ is equal to zero.

Suppose our basic description were in terms of trajectories, as is the case in classical mechanics. Then the transition probability would be one (1) if the domain E would contain the trajectory at point t, and would vanish otherwise. In a genuine probability theory, this is not so. Then, the numbers associated with transition probabilities are positive numbers comprised between 0 and 1. How is this possible? We come immediately to a dilemma. One possibility is to refer to our ignorance: we don't know which trajectory to consider. As a result, we have to give a statistical weight to various possible trajectories. Such an interpretation would make our ignorance responsible for the appearance of probabilities, and ultimately for the introduction of irreversibility in Boltzmann's scheme.

It is difficult, however, to reconcile this interpretation with the constructive role of irreversibility. We know today that irreversibility is at the root of self-organization in chemistry and physics, and plays a central role in biological processes. Therefore, life cannot be the outcome of our own errors, of our ignorance.

The only other possibility which seems open is that for systems to which the second law of thermodynamics applies, the description of reality in terms of trajectories has to be given up. This is obviously a momentous step, and one understands that great scientists such as Einstein have been reluctant to take it.

However, the conflict between fundamental dynamic theories, be it classical dynamics, quantum mechanics or relativity and the second law of thermodynamics, is unavoidable.

In all these fundamental theories entropy is strictly conserved as a result of a general mathematical property which is the *unitary* character of the time evolution. Therefore, it seems that indeed at the fundamental level of description, there exists for classical theoretical physics no place for history, for meaningful changes from order to disorder or vice versa.

P. T. Landsberg discussed this situation in a recent book whose title I find quite appropriate: *The Enigma of Time.* He summarizes some of the positions taken by physicists in the past: for some (probably the majority of physicists) the second law has been regarded as an approximation, or even as anthropomorphic in its character.

I already mentioned why this seems quite unlikely today. For others, irreversibility comes ultimately from cosmology and perhaps from some gravitational correction to be introduced into the equations of motion. This also seems to be quite unlikely. It is true that we are embedded in an expanding universe. However, the second law of thermodynamics is not universal. We may imagine dynamic systems such as the undamped harmonic oscillator or the two-body planetary motion to which we cannot apply the second law.

Still these systems are also embedded in the expanding universe. Moreover, classical dynamics or quantum mechanics have been verified experimentally in simple situations to such a degree of precision that the inclusion of additional terms which would be responsible for thermodynamic irreversibility seems out of question.

For these reasons, we have taken a quite different

approach to the problem of irreversibility. We have taken the law of entropy and therefore the existence of an arrow of time as a fundamental fact. Our task then is to study the fundamental change in the conceptual structure of dynamics which results from the inclusion of irreversibility.

This fundamental change, as we shall see, is precisely related to a revision of the concepts of space and time, whenever irreversibility is involved.

Let us observe that curiously, the two great revolutions in physics over the century have been precisely connected with the inclusion of impossibilities in the frame of physics. In relativity a fundamental role is played by the velocity of light which limits the speed at which we may transmit signals. Similarly Planck's constant h limits the possibilities of measuring simultaneously position and momentum. As noticed by Fritz Rohrlich, "The implications of the finiteness of Planck's constant (h is greater than 0) for the quantum world are as strange as the implications of the finiteness of the speed of light (c is less than infinity) for space and time in relativity theory. Both lead to realities beyond our common experience that cannot be rejected."

In addition to the "impossibilities" which are the result of Planck's constant or of the finiteness of the speed of light, we have the impossibilities which come from irreversibility, the second law of thermodynamics. Only processes which increase entropy in isolated systems are possible. Such a limitation on the macroscopic scale must express also some type of limitation on the microscopic scale. The second law has therefore to appear, as we shall see, as a kind of selection principle propagated by dynamics. The inclusion of this supplementary restriction brings us even further away from the intuitive vision of space and time as used in classical science.

Let us now outline the direction in which we see the solution to the problem of irreversibility.

An unexpected development of modern dynamic systems theory is the importance of unstable systems. Arbitrary small differences in initial conditions are amplified. The situation is represented schematically on the following page.

(See illustration, p. 130, overleaf)

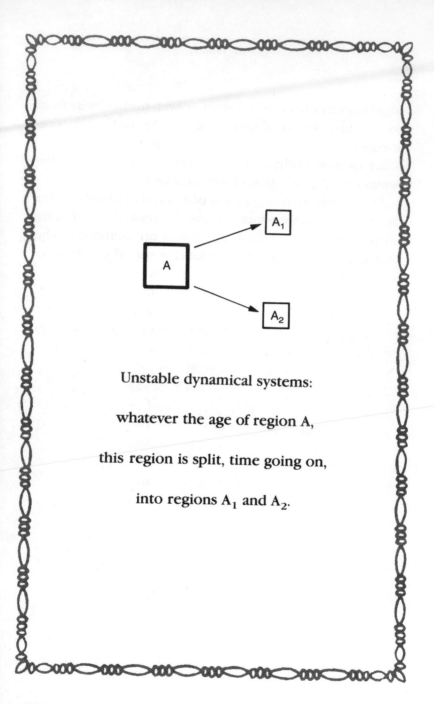

Unstable dynamical systems:

whatever the age of region A,

this region is split, time going on,

into regions A_1 and A_2.

Whatever the size of the initial region A, there are trajectories which lead to regions A_1 or A_2. As each region contains diverging types of trajectories we can no more in a meaningful way perform the transition from finite measure ensembles in phase space (such as region A) to individual points corresponding to trajectories.

Sufficiently strong instability of motion leads to the loss of the concept of a trajectory as a physical meaningful concept. This is a fundamental fact which makes possible the incorporation of probability and irreversibility in physical theory without invoking the idea of ignorance.

A simple example of an unstable dynamic system is provided by the so-called baker transformation which we shall use again later as an example. It may be seen as the transformation B of the unit square onto itself which is the result of two successive operations: (1) first the unit square is squeezed in the vertical direction to half its width and is at the same time elongated in horizontal directions to double the length; (2) next, the resulting rectangle is cut in the middle and the right half is stacked on the left half. The iterates of B may be considered to model the dynamic evolution of a system at unit interval of time.

(See illustration, p. 132, overleaf)

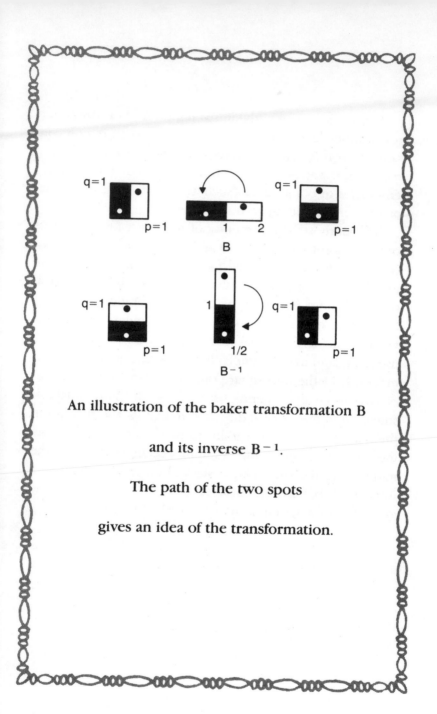

An illustration of the baker transformation B

and its inverse B^{-1}.

The path of the two spots

gives an idea of the transformation.

A basic feature of highly unstable systems, which was recognized by B. Misra, is that we may introduce for such systems a new concept, corresponding to the "internal time" or "internal age." Internal time is quite different from the usual parameter time, which I can read on my watch.

It corresponds more closely to the question which I ask when I meet a stranger and I wonder how old he is. Obviously the answer will depend on the overall appearance. His age cannot be read from the colour of the hair, the wrinkles on the skin. It depends on the global aspect. Internal time comes closer to ideas recently put forward by geographers, who have introduced the concept of "chrono-geography."

When we look at the structure of a town, or of a landscape, we see temporal elements interacting and coexisting. Brasilia or Pompeii would correspond to a well-defined internal age. On the contrary, modern Rome, whose buildings originated in quite different periods, would correspond to an *average* internal time.

For simple unstable systems such as those corresponding to the baker transformation, illustrated above, we may reach a more quantitative understanding of internal time.

Let X_0 be the function which assumes the value -1 on the left half of the square and $+1$ on the right half. Let us define $X_n = U^n X_0$ corresponding to the application of n baker transformations. A few of these iterations are represented on the following page.

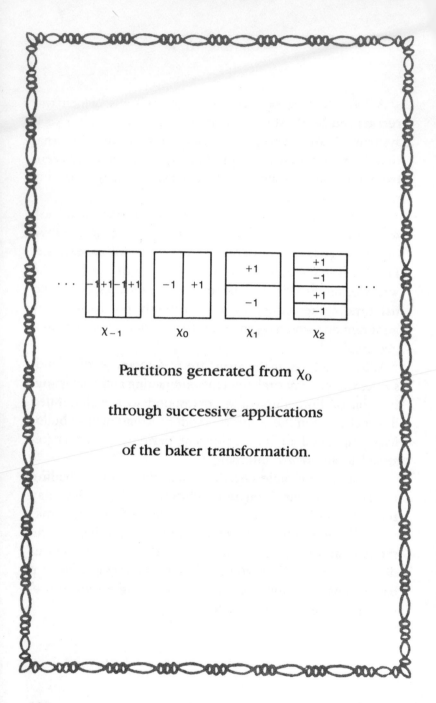

Partitions generated from χ_0

through successive applications

of the baker transformation.

The various functions X_n are "eigenfunctions" of internal time. The internal time is necessarily an *operator* like the one we use in quantum mechanics.

Arbitrary partitions of the square do not have a well-defined internal time, but only an average internal time.

In contrast, the partitions represented on the illustration just provided correspond to well defined internal times starting with o for partition X_0. The age of the partition X_n is the number n of iterations I have to perform to go from X_0 to X_n.

Whenever the internal time exists it is an *operator,* and not a number. It is important to grasp this difference: an arbitrary partition of the square has no well-defined internal time (as has the partition X_n). In general, we can only speak of an average internal time.

Instead of using the baker transformation to illustrate these ideas, we could use a glass of water into which we pour a drop of ink. The internal time is now related to the shape the ink takes; but an arbitrary distribution of ink in water has no well-defined internal time, as the ink may have been introduced at various times.

The existence of an internal time operator has some far-reaching consequences. We now are able to describe the evolution of the system in terms no more of trajectories, but of partitions. Obviously, these two descriptions, one in terms of partitions, the other in terms of trajectories, are complementary in the sense used in quantum mechanics (to describe, however, a physically quite different situation). If the state is described by a partition, we know only that the system is in a region of phase space; but we don't know its exact location. Similarly, a point in phase space may be embedded in an infinite number of partitions. The internal age of a trajectory is undefined.

In more technical terms, the dynamics of unstable systems equipped with internal time corresponds to an algebra of noncommuting observables. Once we use internal time and partitions we have lost the local point of view of classical mechanics. Instability leads to *non-locality.* In this way, the main obstacle for the transition between dynamic theories and probabilistic description is eliminated. As long as the basic description used in classical mechanics was the trajectory, there was no hope to reach a microscopic theory of irreversible processes. But for highly unstable dynamic systems we have an alternative way, which involves a topological description, and eliminates the appeal to trajectories.

It is only for these systems that the second law of thermodynamics may be meaningful in an intrinsic sense, and not be the mere outcome of approximations or errors. These systems are of tremendous importance, as they encompass all of chemical systems, and therefore also all of biological ones.

We have now reached the core of the problem: What is time? According to Carnap:

> Once Einstein said that the problem of the Now worried him seriously. He explained that the experience of the Now means something special for man, something essentially different from the past and the future, but that this important difference does not and cannot occur within physics.

That this experience cannot be grasped by science seems to him a matter of painful but inevitable resignation. I remarked that all that occurs objectively can be described in science: on the one hand the temporal sequence of events is described in physics; and, on the other hand, the peculiarities of man's experiences with respect to time, including his different attitude toward past, present and future, can be described and (in principle) explained in psychology. But Einstein thought that scientific descriptions cannot possibly satisfy our human needs; that there is something essential about the Now which is just outside of the realm of science.

As I mentioned earlier, we begin to see a way out of the difficulty which plagued Einstein. But the concept of time which may incorporate the "Now" in a more fundamental sense is indeed quite different from the traditional, linear representation as it came to us from Aristotle.

We could in fact imagine a world in which we would not age all together: the future of some would be the past of others. This is, however, not our world. As we have seen for unstable dynamic systems, for which we can define the internal time, a different description becomes available. As an example, consider a distribution in phase space as represented in the illustration to be found on the following page. We can describe this distribution as a superposition of the basic partitions introduced in the illustration given immediately before this one.

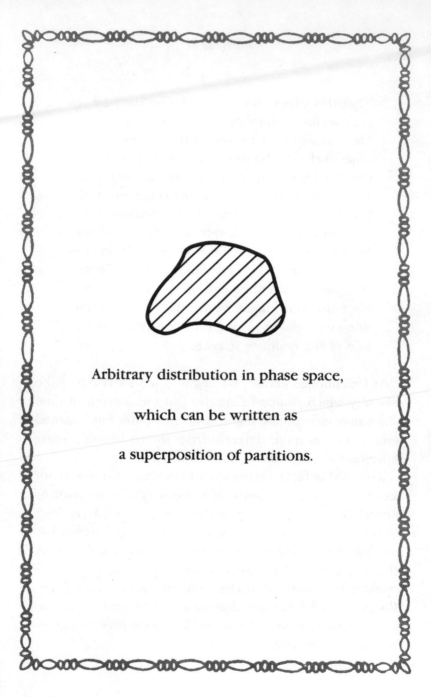

Arbitrary distribution in phase space,

which can be written as

a superposition of partitions.

In mathematical terms, this corresponds to the formula:

$$\zeta = \sum_{n = -\infty}^{n = +\infty} c_n X_n \qquad \text{(Equation 1)}$$

The index $n = 0$ corresponds to the present; the values "n is greater than 0" correspond to the future, while the values "n is less than 0" correspond to the past. The important point is to notice that Σ extends symmetrically over the past and the future. X_n is the partition corresponding to internal time n.

This confronts us with a quite interesting situation: while the classical distribution of past, present, and future refers to a given, "astronomical" time (time as read on a watch), the new description, as expressed in the mathematical formula just given, combines contributions coming from all values of the internal time. In this sense, time becomes "non-local": present is a recapitulation of the past, and an anticipation of the future.

A comparison with our own situation may help. The present state of our neuronal system contains an essential element of our past experience, and an element of anticipation of future events. However, for time as it is implied for example in neuro-physiological activity, present and past cannot appear as symmetrical.

We may now introduce this asymmetry, or, equivalently, the second law in our description. Basically, this corresponds to giving a different weight to the past and to the future, as the following formulas and equations will show:

Instead of the distribution function ζ we now introduce an appropriate transform of ζ which we shall call $\tilde{\zeta}$ and which can be shown to satisfy a probabilistic evolution equation, and reach equilibrium for the distant future:

$$\tilde{\zeta} = \Lambda \, \zeta \qquad \text{(Equation 2)}$$

In this formula, Λ may be constructed when the internal time T is known. In fact, it is a decreasing function of the internal time. Instead of (Equation 1), we may now write:

$$\tilde{\zeta} = \sum_{-\infty}^{+\infty} c_n \, (\lambda_n \, X_n) \qquad \text{(Equation 3)}$$

Again, the Σ extends from $-\infty$ (the far distant past) to $+\infty$ (the far-distant future). But there is an essential difference with (Equation 1). The contribution X_n corresponding to internal time n is multipled by a number λ_n (the value of Λ for T = n).

The numbers λ_n are positive, and form a decreasing sequence, as (Equation 4) will show:

$$\lambda_n \to 0 \quad \text{For} \quad n \to +\infty \qquad \text{(Equation 4)}$$

This has an important meaning: future is open from the point of view of internal time. Indeed, the contributions coming from n positive and large are damped by the multiplication with λ_n.

In other words, future is not contained in the present for

systems satisfying the second law of thermodynamics. Therefore, according to this description, states have an orientation in time. Time is now intrinsic to objects. It is no more a container of static, passive matter.

I find it quite striking that the closest links with the conclusions we have reached are to be found in the work of two poets.

One is Paul Valéry; let me quote one of the remarks we find in the *Cahiers:*

> —En somme, je *crois* qu'il y a une mécanique mentale qu'il ne serait pas impossible de préciser.
>
> Mais cette mécanique, qui doit s'inspirer de l'autre toutefois ne doit pas craindre de prendre ses libertés nécessaires — c'est-à-dire de contredire la première sur les points qu'il faut.
>
> Ainsi la varible temps est profondement differente. Le temps mental est plus une fonction qu'une variable, en psychologie — et on trouvera $\frac{t}{F}$ plus souvent que $\frac{F}{t}$.

This is a most vivid evocation of the topological time we have been describing in this lecture.

The other is T. S. Eliot. You know these verses from "Burnt Norton":

> Time present and time past
> Are both perhaps present in time future
> And time future contained in time past

It would be difficult to express in a clearer way the connection which exists between past, present and future. But Eliot continues:

> If all time is eternally present
> All time is unredeemable

Indeed, time would be unredeemable in a deterministic universe. But in a universe submitted to the second law whose microscopic foundations imply instability and therefore a stochastic description of time evolution, time is redeemable. As a result, we begin to understand the difference between the tautological universe, which has obsessed us since the dawn of physical thought, and the reality of time we experience in the world in which we have been thrown.

We have been led to the conclusion that broken time-symmetry is an essential element in our understanding of nature. A simple musical experiment may illustrate what we mean by this statement. We may play a sound sequence during a given time-interval, say, one second, starting for example with *piano* and ending with *fortissimo*. We may play the same sequence in reverse order. Obviously, the acoustical impression is deeply different.

This can only mean that we, equipped with an internal arrow of time, distinguish between these two performances. In the perspective which we have summarized in this lecture, this arrow of time does not oppose man with nature. Far from that, it stresses the embedding of mankind in the evolutionary universe which we discover at all levels of description.

Time is not only an essential ingredient of our internal experience and the key to the understanding of human history both at the individual and at the social level. It is also the key to our understanding of nature.

Science, in the modern sense, is now three centuries old. We may distinguish two moments where science has led us to a well-defined image of the nature of physical existence:

—One was the moment of Newton, with his world view formed by changeless substances and states of motion, with a conception in which matter, space and time were dissociated as time and space appeared as passive containers of matter.

—A second state was reached by Einstein. Pehaps the greatest achievement of general relativity is that space-time is no more independent of matter. It is itself generated from matter. Still, in Einstein's view, it was essential to keep the idea of localization in space-time as an integral part of the theory.

We now begin to reach a third stage, in which this localization in space-time is submitted to a more thorough analysis. Curiously, this questioning of the microscopic structure of space-time emerges at present from two quite independent directions: quantum theory and the microscopic theory of irreversibility.

Our relation with nature, and especially the problem of learning and measuring, become only meaningful in this perspective, which incorporates instability and irreversibility.

It is remarkable to see how close some recent conclusions

are to the anticipations of Whitehead and Heidegger.

In his basic work *Process and Reality,* Whitehead emphasizes that simple location in space-time cannot be sufficient, that the embedding of matter in a stream of influence is essential. Whitehead emphasizes that no entities, no states, can be defined without activity. No passive matter can lead to a creative universe.

The title of Heidegger's influential book, *Being and Time,* is in itself a manifesto, emphasizing Heidegger's opposition to the timeless concept of Being, which corresponds to the mainstream of western philosophy since Plato.

States may be associated with Being, and time evolution with Becoming. States as defined by Equation 1 are time-symmetrical (in reference to internal time).

We now have an example of the relation between Being and Becoming, as often described in western physics. Being is independent of Time. But this description does not include the second law of thermodynamics. Once this is done, we come to relations expressed in Equations 1 and 2 with a broken time-symmetry, which is then propagated by time-symmetry broken laws of evolution, including the second law of thermodynamics.

From a logical point of view, there are therefore at least four possible solutions to the problem of Being and Becoming. However, our existential situation allows us only to retain the solution involving a broken time symmetry.

Two centuries ago, Kant asked three questions: What can I know? What should I do? What may I hope? He thought that only speculative philosophy could give contributions to the answers. I believe today the situation appears as quite different. Science can also give a contribution to the basic interrogations of mankind.

We have overcome the basic duality between man and the universe: Time was the main element in the opposition between man and the universe.

It seems to me that we are living in a most exciting moment of the history of science. We slowly come to a description of time which, in addition to its traditional distinctive features, incorporates some of its main connotations, such as irreversibility, evolution and creativity. This century has already known two great revolutions in basic theoretical physics. Whatever the detailed methods will be, it seems clear to me that we approach a point where the rediscovery of time will lead not only to a better understanding of the mechanisms of change, which we encounter on all levels of the universe we observe, but also to a better embedding of man in the universe from which he has emerged.

As beautifully summarized by G. Steiner in his comment on Heidegger, "the human person and self-consciousness are not the center, the assessors of existence. Man is only a privileged listener and respondent to existence."

The new description of time puts in a new perspective the question of the ethical value of science. This question could have no meaning in a world viewed as an automaton. It acquires a meaning in a vision in which time is a construction in which we all participate.

Sudden renewal of the self—from where?
A raw ghost drinks the fluid in my spine;
I know I love, yet know not where I am;
I paw the dark, the shifting midnight air.
Will the self, lost, be found again? In form?
I walk the night to keep my five wits warm.

Dry bones! Dry bones! I find my loving heart,
Illumination brought to such a pitch
I see the rubblestones begin to stretch
As if reality had split apart
And the whole motion of the soul lay bare:
I find that love, and I am everywhere.

From "The Renewal,"
by Theodore Roethke

CONVERSATION THREE

&

THE FUTURE OF HUMANS IS
THE HUMANS OF THE FUTURE

&

A FINAL EXCHANGE BETWEEN
ROGER SPERRY, SIR JOHN ECCLES
AND BRIAN JOSEPHSON

ABOUT CONVERSATION THREE

One revolutionary repercussion that has emerged thus far in these Conversations at the Isthmus Institute in Dallas is the challenge to reinterpret the relation between nature and humankind. Our Nobel Laureates, each from his own unique perspective, have boldly questioned some of science's most sacrosanct conventions, especially its predilection for materialist reductionism. Dr. Roger Sperry and Sir

John Eccles have confronted us with facts that demonstrate the enfolding power of non-material mind vis-à-vis material brain. Dr. Sperry, especially, has insisted upon science's need to revise its understanding of cause and effect through his proposal regarding a "downward causation" that significantly alters scientific assessments of the relation between vital higher-level forces (such as mind) and lower-level ones. Dr. Brian Josephson has extended this notion further still by inviting consideration of a science which proceeds by investigating the "very top," including "God" ("Mind," "Supreme Intelligence") as the enfolding force which sustains the presence of an implicate, unobserved order within the evolving universe. Dr. Josephson does not, however, reach his position through "downward causation." And in Interlude II, just ended, Dr. Ilya Prigogine challenged science to rediscover time by tracing the foundations for an irreversible arrow of time which is at work on both microscopic and macroscopic levels. Prigogine's description of time as "construction" in which all participate overcomes the basic duality between humankind and the universe and establishes the basis for a new dialogue with nature. As he noted, "we approach a point where the rediscovery of time will lead not only to a better understanding of the mechanisms of change, which we encounter on all levels of the universe we observe, but also to a better embedding of man in the universe from which he has emerged."

While we cannot claim that these revolutionary challenges constitute a unified concept of the relation between human life and nature, they do converge upon one common concern: the human future and future humans. Each of these scientists is interested in understanding the uniqueness of human existence in an evolving universe, while at the same time stressing the intrinsic relatedness between humans and the world of nature in which they participate. All of these Nobel Prize winners envision a future universe within which humans are entirely at home as both comfortable participants and avid observers. Their search for a renewal and revision of science is simultaneously a quest for human renewal.

It seems quite appropriate, therefore, that this third and final Conversation, involving Roger Sperry, Sir John Eccles and Brian Josephson, should open with stanzas from Theodore Roethke's poem "The Renewal." In Roethke's poetic imagination, the quest for personal renewal of the self becomes metaphor of a more universal seeking and a longer, more arduous journey toward the heart of reality. In travelling a personal road to find "my loving heart," Roethke intimates, each human receives illumination of the whole.

> Illumination brought to such a pitch
> I see the rubblestones begin to stretch
> As if reality had split apart
> And the whole motion of the soul lay bare . . .

The "I" of this poem is a pilgrim whose own renewal, from "dry bones" to heart of flesh, presages a renewal of life within all reality, when the "whole motion of the soul" is awakened by the discovery that "I am everywhere."

Roethke's "I" thus seems very close to that vital force of mind enfolding brain, time and matter. Of enormous importance in this last Conversation is a shared conviction that the human future is one we can choose, one in which we can discover meaning, mystery, purpose and, indeed, love. The Conversation is particularly notable because it lacks the foreboding images that so often festoon "futuristic" writing and popular science. Here one finds no bald women in Mercury-boots, no university professors reverting to simian grunts and leaps, no mythical Martians demanding to be taken to our leader. The future of the evolving universe envisioned by the participants in this third Conversation is recognizably human and humane.

Sir John Eccles initiates this Conversation by raising perennially critical questions of personal freedom and moral responsibility. From his own research into the relation between mental intentions and neural events, he argues strongly in behalf of both freedom to know and freedom to act, to develop ourselves as persons, to fulfill ourselves through responsible interaction with family, society, loved ones and friends. Dr. Roger Sperry then picks up the theme, extending it to a discussion of how, in future, the relation between science, religion and the humanities might be perceived—as these relate to the search for new values and ethical principles. This vision of the human future is developed still further by Dr. Brian Josephson's proposals for a "science of mysticism" for which a mathematical basis might be established.

In the estimation of these scientists, the evolving future of our universe need not be one in which human beings are diminished or lost. It need not be inevitable that the quality of human life steadily deteriorate in a polluted, over-populated biosphere or collapse utterly through a nuclear Armageddon. The research of Sperry and Eccles into the enfolding potency of mind in relation to brain reveals that the fundamental human powers of choice, decision and freedom to act can, in the continuing course of evolution, become ever more replete with purposefulness and meaning. From the perspective of evolving time, the distinctively human face of our universe is only just now beginning to appear.

Conversation Three opens with the much-debated question of freely willed action as it may apply to complex social and moral situations. The first speaker is Sir John Eccles.

SIR JOHN ECCLES

It can be appreciated that the demonstration of a free-will action is much more convincing when a very simple intention or attention is under consideration than when a human action is studied in some complex physical or social or moral situation. In the first place it is essential to recognize that most of our actions are learnt repertoires of skills which we carry out automatically without mental concentration or even without conscious awareness, for example, driving a car in light traffic on easy, well-known roads.

Mental attention and decision are only used in an emergency. Secondly, philosophers tend to speculate on freedom of action in some complex social or moral context. For example, if we meet a visiting professor in the street, should we invite him home for dinner? I would not contest the conjecture that such a decision could be freely made, but the difference between this complex situation and that of finger flexion has an analogy with the attempt to discover the laws of motion by analyzing the eddying movements of a turbulent river instead of the Galilean method with mental balls rolling down an inclined plane!

If we can establish that we have freedom to bring about simple movements at will, then more complex social and moral situations must also in part at least be open to control by a voluntary decision, i.e., of mental thought processes. Thus we have opened the way to the consideration of personal freedom and of moral responsibility.

Though theoretically we may be free to choose between alternative responses to a situation, this choice may be vitiated by constraints. In a totalitarian state these constraints are enforced by severe penalties. How far then can these unfortunate citizens be held morally responsible for their decisions and actions? Only the brave dare to challenge the power of the state by dissent and to suffer the inhuman consequences. We may ask: How far can the people be held morally responsible for actions done under such constraints? At least we can realize that human freedom is vitiated by constraints applied by a police state. Yet freedom is also corroded in a society that is too permissive of violence and crime.

The freedom that matters is the freedom to know, freedom of thought, of opinion, of discussion. Such a freedom does not limit the freedom of others, and there can be no doubt that it is fundamental, but as well as "freedom to know" we also want freedom in the sphere of action. Freedom to make by our own efforts the utmost out of our lives, to develop ourselves as persons, to give our talents full scope, to live according to our ideals, to control our destiny. This is the freedom that Maritain aptly calls "freedom in fulfillment." This freedom, too, does not limit the freedom of others. Moreover, the freedom of fulfillment obviously includes the "freedom to know." We can recognize that fulfillment means a full life in the family and in society with loved ones and friends associated in organizations for culture and worship and also for our work and recreation.

How can we set about building our society so that we preserve not only the freedoms we already have, but add to them so as to give the fullest possible life to all persons, so creating an order in which all the varying richness of the human personality will be manifested? That should be the central political problem of this age. It is not sufficient, however, to provide such opportunities. Each person should, by his own will, strive to make the most of these opportunities. Freedom involves not only rights, but also duties. We have no unqualified right to freedom. We are only entitled to freedom in so far as we fulfill the duties of respecting and living up to the freedom we already have. For example, freedom of fulfillment necessitates a progressive conquest of the fullness of personal life and spiritual liberty. Thus, it is evident that we can never attain a static state of freedom. Freedom is dynamic in the sense that we have to be continually striving for it in order even to maintain what we have. Abuse of our freedom endangers it. For example, we endanger our freedom of speech when we use it to make misleading, irresponsible and provocative statements. The right of freedom of speech presupposes the duty of honesty and sincerity.

Our world is at the parting of the ways. There are only two alternative orders towards which we can move.

One way leads to the centralized planning of the absolutist slave State. In the past a private and personal sphere of life was often able to survive almost untouched by an absolutist tyranny, for despotism was largely devoted to public affairs. With modern efficiency of communication and organization that is no longer possible. Absolutist governments can only stabilize themselves if they eliminate all resistance before it can organize. Secret police, concentration camps, treason trials and mass propaganda are the inevitable concomitants. Modern absolutism must be total, enslaving man even in his personal and private life. It must be a tyranny characterized by compulsion, terror and collectivization of every aspect of life. Such slave States may be very stable.

The other way leads by continuous development to the order of freedom and moral responsibility of each human person. It will be an order respecting and nurturing the private lives of all persons, and dependent dynamically on the responsible and free acts of each one of them. At every level of society there will be full play for responsible action. In other words, one of the first considerations will be the widest possible extension of personal responsibility. It is the freely exercised moral choice which most fully expresses personhood and best deserves the great name of freedom.

We can say that it is of transcendent importance to recognize that by taking thought we can influence the operation of the neural mechanisms of the brain. In that way we can bring about changes in the world for good or for ill. A simple metaphor is that our conscious self is in the driver's seat. Our whole life can be considered as being made up of successive patterns of choices that could lead to the feeling of fulfillment with the attendant happiness that comes to a life centered on meaning and purpose.

Each human person is a unique self with potentialities that give wonderful promise with all their great diversity. The ideal is for each human person to have the maximum freedom to realize its potentialities. This ethic derives from the belief that life has a transcendental meaning and that each life is precious. Together we are, as human persons, engaged in the tremendous adventure of consciously co-experienced existence.

COMMENTARY

Two convictions stand out prominently in what Sir John has just said: that personhood expresses itself most fully in freely exercised moral choice, and that the precious quality of each individual life is enhanced by its social character, by the "tremendous adventure of consciously co-experienced existence." His view is thus distinguished from simplistic political philosophies of both the right and the left which believe either that individual integrity can be maintained only by sacrificing larger aspects of the social welfare, or that "the common welfare of all" demands systematic suppression of the individual. Individual "freedom of fulfillment" and socially co-experienced existence are mutually re-enforcing potentials. Together, they reveal the possibility of a transcendental meaning for human life—a point of view quite congruent with Theodore Roethke's "The Renewal," which argues that the personal quest for love succeeds only when it leads to reaching beyond the self to touch and be touched by that world which is wider than the philosophy of materialism so often accepted by science.

DR. SPERRY

I agree. Until very recently, the acceptance of science has meant embracing the philosophy of materialism along with the interpretations of human nature and society which this implies. Marxism upholds values and a worldview that are substantially opposed to the ones that would emerge from a system based on science as we here understand it. In Marxism, what counts in shaping the world and human affairs are the actions man takes to fulfill his material needs. But this overlooks the key principle of downward causation. Under the mentalist view, traditional reductionist interpretations emphasizing control from below upward are replaced by revised concepts that emphasize control from above downward. The higher idealistic properties that have evolved in man and society can supersede and control and take care of the more primitive needs.

The espousal of science by the Marxists and many others, including the secular Humanists, has usually meant also the rejection of institutional religion. This seems a mistake, especially with world conditions as they are. More than ever there is need today to raise our sights to higher values above those of material self-interest, economic gain, politics, production power, daily needs for personal subsistence, etc., to higher, more long term, more godlike priorities.

What this recently revised outlook in science might mean for a merger with religion, and for the kind of value-belief system, ethic and theology that might emerge has yet to be developed. Concepts of salvation, transcendent meaning, ultimate value and such like would have to be redefined and translated into a reference frame consistent with the worldview of science. The task can be likened in some respects to that of trying to deduce what form religion and the teachings of Christ, Muhammad, Buddha, Confucius, and other founders, might have taken, if Copernicus, Darwin, Einstein, and all the rest had come before their time instead of after. It is something that would take time to develop and many volumes to describe in full, with separate books for each religious view and denominational variation. A long effort over some two decades has been made in this direction for Christianity by Ralph Burhoe and his associates with their journal *Zygon* and the *Institute on Religion in an Age of Science.* But, of course, the general idea of bringing religious belief into harmony with scientific reality is centuries old and widely apparent in liberal theologies.

From the standpoint of science, one can foresee at least a few broad generalities that derive from the constraints set by science, and would seem to apply in common across the board to any value-belief system or theology derived on our current terms. As already mentioned a central requirement imposed by science would seem to be a relinquishment of dualist concepts in conformance with the explanation of mind in monist-mentalist terms. Such a shift from various dualistic, otherworldly beliefs to a monistic, this-world faith, would mean that our planet should no longer be conceived, or treated, as merely a way-station to something better beyond. This present world and life would thus in each case, acquire an added relative value and meaning.

Scientific doctrine regarding evolution, causation, and the current concepts of emergent forces and downward control would also appear to exclude any distinct separation of evolving creation from the intrinsic creative forces or force system. In this sense, science supports Spinoza's contention that the Creator and Creation cannot be separated. The two of necessity become intimately interfused and evolve together in a relation of mutual interdependence. Thus, what destroys, degrades or enhances one does the same to the other. Creation itself therefore, i.e. all evolving nature including the human brain and human psyche, logically takes on a relative degree of sacredness not present in thinking where the things that are most sacred are set apart in another form of existence.

COMMENTARY

By drawing out some of the implications that result from science's mentalist revolution and its corollary of downward causation, you indicate, Dr. Sperry, that dualist understandings of the relation between Creator and Creation are inappropriate. It may be of interest to note that, in the estimation of a great many modern theologians, the best Jewish and Christian thinking on this subject has similarly repudiated dualist perceptions that pit "profane" earthly realities against other-worldly, sacred ones. As modern

*scholars reconstruct it, for example, the preaching of the historical
Jesus (as distinct from the Christ reflected throughout the faith of
later New Testament literature) did not appeal to a "heavenly
zone" of existence separate from and superior to earthly life. At the
center of Jesus' message, it seems, was the reign or kingdom of God,
not a place (a "heaven"), nor an object of reward, but an activity
of God within this world on behalf of this world's human beings.
Jesus' God, the God of Israel's faith, was one intent on humankind. For Jesus, then, God's "cause" was, quite simply, the human
cause. God becomes, in other words, the one for whom all things
make a difference and who makes a difference to all things. God
could perhaps be described, in this view, as a center of interaction
—of action on and reaction to—all other things. And if that is the
case, this God would "become," and be experienced, within an
evolving universe which moves, as you noted in Conversation Two,
Dr. Sperry, toward increasing complexity, beauty, meaning and
purpose.*

*There are, of course, still further implications that would emerge
from the "this-world reality" demanded by a science revised according to mentalist terms.*

DR. SPERRY

When we relinquish authoritarian, otherworldly criteria
and make values referent to this-world reality in accordance
with the worldview of science, values are no longer absolute or infallible, though some aspects of reality are relatively constant. If reality changes, however, as it has in
respect to human numbers, ethical and moral values also
change. Even the sanctity of human life is not immune, does
not fully escape the laws of mathematics or of supply and

demand, nor the demeaning effects of excessiveness. Over-population becomes doubly immoral, not only because of the effects on the biosphere in general, but also because of the effects on the quality, value and meaning of human life itself. We customarily recognize a kind of beauty and added worth in rarity and vice versa. The growing sense of value-lessness and meaninglessness in modern society can be correlated in no small degree with the very real increased expendability and anonymity of the individual caused by today's overwhelming numbers.

Human nature evolved in small communities where individuals counted, heroic leaders were possible, contrasts were everywhere, and life was in close harmony with nature. When we compare this with today's faceless hordes of massed humanity struggling for what is often a socially meaningless existence in the larger overcrowded cities of our world, one has to wonder if something isn't morally very wrong. Trials and degradation in this life may not matter so much if there is an eternal hereafter to look forward to, but in the absence of such futures, this-world reality becomes a much greater concern. It is along the foregoing and related lines that the current revisions in the worldview of science, when merged with theology, are seen to lead to value perspectives that make it immoral, even sacrilegious to pollute, to overpopulate, to waste irreplaceable resources, to carelessly exterminate other species or in any other way to destroy, degrade, or desecrate the quality of the biosphere for coming generations.

Many religious believers hold that it is impossible to join religion and science on the terms described above without seriously undermining or destroying religion. To have to give up beliefs in a personal deity that is omniscient and caring, or belief in an immortal soul that survives bodily death along with the kind of added purpose and life meaning these endorse, seems for some people like having to give up the very essence and central core of religious faith. It is argued that such dualistic beliefs satisfy deep emotional needs in a way that a 'scientific theology' never can and that mankind throughout history has universally in all cultures depended on otherworldly spiritual beliefs of this kind.

In partial answer, one can point to recognized religions that lack a personal deity and to deeply religious persons, including religious leaders, who have conceived of God in untraditional ways. One can also point to the many 'nonbelievers' of today, to the Communist world, to the secular Humanists, agnostics and adherents of 'liberal' faiths that collectively make up a substantial fraction, if not the majority, of the world population. It has already been mentioned that the scientific view of man's creator, perceived in mentalist terms, need not be strictly impersonal, purposeless and uncaring, as was the case with reductive scientific materialism. From the viewpoint of the human species as a whole one may think of evolving nature in impersonal terms, especially if cultural evolution is omitted, but from the personal standpoint of the individual the perspectives become quite different. When it comes to the individual personal perspective, the parents and ancestors obviously have to loom very large among the forces of creation. So also do other family members, friends, teachers and the whole community of people by whom the individual is influenced and who thereby help to create the kind of person one becomes. In adulthood, one's mate and other intimate relations have to be included among the important movers and shapers of the human psyche.

In other words, the importance of religion fulfilling personal emotional needs and life meaning of this kind would not need to be deemphasized or lost but only retargeted into this-world reality. With public faith oriented in this direction, one could expect relevant changes in the structure and institutions of religion and society that would make them better suited to handle these kinds of needs. The current success of cults like the Hare Krishna and the Moonies and others is probably not based so much on anything distinctive about their other-worldly doctrine as upon their this-world practices that use "togetherness," communal effort and related things that help fill unsatisfied psychological needs.

COMMENTARY

The view you are outlining here, Dr. Sperry, seems quite close to one which many sociologists of religion label as "functionalist." This functionalist interpretation sees religion's primary concern as meeting deep-seated human needs, thereby creating emotional equilibrium and satisfaction. This interpretation has often been used of dualist, otherworldly belief systems, but it need not be restricted to them and, as you have implied, can be invoked just as easily to support a this-world faith more in line with the new science described in these Conversations.

Still, the functionalist interpretation appears to ignore some fundamentally dysfunctional, yet important, aspects of religious faith and experience. Religion, as is well known, may not only sustain emotional equilibrium, it may create far-from-equilibrium conditions as well. Historically, religion has served not only to reconcile believers to their experience and integrate them into society, but to create friction between belief and culture and even to produce internal dissatisfaction within the believer. For both individuals and societies, religion may act as a disruptive force leading to counter-cultural militancy or to profound personal shifts in values and actions (as happens, for example, in "conversion.") Some might feel, Dr. Sperry, that your view simply applies the reductionist position which you oppose in science to the phenomenon of religion. Or it might be argued that in relating your theory of downward causation to religion you have not started quite far enough at the top—that among the mental forces and intentions which enfold and supersede lower-level entities, one should include a Mind who supremely typifies intelligence, mental intention and choice.

There is, further, a broad range of religious experiences, some of which were touched upon briefly by Dr. Brian Josephson in Conversation Two, which seem to escape sociological categorizations altogether. Here, for example, is a poem by Emily Dickinson:

> He fumbles at your Soul
> As Players at the Keys
> Before they drop full Music on—
> He stuns you by degrees—
> Prepares your brittle Nature
> For the Ethereal Blow
> By fainter Hammers—further heard—

Then nearer—Then so slow
Your Breath has time to straighten—
Your Brain—to bubble Cool—
Deals—One—imperial—Thunderbolt—
That scalps your naked Soul—

When Winds take Forests in their Paws—
The Universe—is still—

The experiences of "God" and world projected by this poem certainly seem real enough, but Dickinson has skillfully subverted most of the conventional categories linked with either functional or dysfunctional interpretations of religion. Perhaps it is possible that valid religious experiences exist which science, even in revised mentalist form, is not yet fully equipped to explain.

DR. SPERRY

Perhaps that is so. Historical and related humanistic truths and concepts may often be as valid and important in creating modern civilized man as are those double-checked by science. Strict separations between science and the humanities, between fact and value, do not hold as they used to in materialist thinking. Valid insights contributed from the humanities have to be included. What counts is validity. Science is emphasized because of its rigorous standards for validation. Also, science, like revelation, takes us beyond the bounds of ordinary experience. Science gives deeper insights into the nature and meaning of things. It helps clear the mystery and show the way. It enables us to get a better and more intimate understanding of the forces that made, move and control the universe and created man.

Along with the higher human factors, the scientific view includes also, of course, the cosmic and the subatomic and everything in between—the grand overall design of the evolving web of creation of which we are each a part, and the whole matrix of multinested inner forces and energies involved.

Doubts about the possibility of joining science and religion are usually strongest in respect to "after-life" concerns. This is where the conflicts are most acute and seemingly irreconcilable and where it is most difficult for science to compete with dualist faiths in fulfilling related emotional and psychological needs. Everything in science to date seems to indicate that conscious awareness is a property of the living functioning brain and inseparable from it. The conclusion from mind-brain science seems inescapable: that the conscious self, as we ordinarily experience it, does not survive brain death.

Despite the seemingly discouraging prospects of the scientific position, there are some pluses to consider, a few of which seem appropriate to mention because of relevance to our present argument. As pointed out by Popper among others, death adds greatly to the meaning and value of life. What illness does for the appreciation of health, death does for life. Conversely the depreciation of this life and this world by the assumption of a "better beyond" and an "eternal hereafter" leads to a "way-station" perspective on life that is degrading to the most sacred gift the universe offers.

When the many related pros and cons are weighed concerning the alternatives of a world with, and a world without death, the balance appears to come out very heavily in favor of nature's having made the right choice. The prospect of a biosphere without death is to science a contradiction in terms and irreconcilable with evolution and the creation of man.

Varying traditional versions of what aspect or form of the human psyche can survive brain death are, of course, numerous and tend to be vague and conflicting. If we start from scratch and ask in the light of modern knowledge what aspect of the conscious self would be best to preserve, from the standpoint of cosmic design and all things considered, the possibilities allowable by current mind-brain theory are not all negative. In fact, if the aim is to capture and preserve beyond brain death the conscious Self in its very highest form, then an argument can be made that this is, in a sense, provided for in realistic terms in the new mentalist view of the mind-brain relation. The most important thing about the human psyche in this view is not the atomic, molecular, or physiologic infrastructure but rather the supersedent mental events, forces and properties, per se. When it comes to selecting the best of the mental experiences, in the sense of the most highly evolved, there is reason to think that the best is not represented among the everyday thoughts, feelings, wants, fulfillments and other common experiences associated with bodily subsistence and welfare. One looks rather to the higher special peaks in the mental life, and not to the living neural substrate of these but to the transcendent mental content itself that emerges at the very top of the multinested neuro-molecular-atomic-subatomic brain hierarchy. On such terms one can then infer that perhaps the essence of the very best of the conscious self of Beethoven, of Shakespeare, Michelangelo, etc., are still with us. We can't all be Beethovens, of course, or Leonardos, or Edisons, or Darwins, etc., but there are ways in which the highest aspect or form of the conscious experiences of each individual can realistically be extended in this manner to exist beyond death of the neural substrate that originally sustained it.

COMMENTARY

Earlier in this Conversation, Dr. Sperry, a certain convergence was noted between your criticisms of dualist, otherworldly belief-systems and some of the more recent thinking of Jewish and Christian scholars about the meaning of God and of God's relation to the world, especially in the teaching of figures like Jesus. The comments you have just made about joining science and religion with respect to after-life concerns, as well as your assessment of what might "survive" of us following brain death, shows still further grounds for dialogue. A number of modern Christian theologians have been offering re-interpretations of such beliefs as "'resurrection" and "immortality." Until the very recent past, these two beliefs were interpreted in primarily biological categories: resurrection meant "resuscitation," restoration to a former flesh-and-blood existence, while immortality was conceived of as survival of the self-conscious, rational identity. Similarly, the resurrection of Jesus was thought to be a restoration similar to the events surrounding the coming-back-to-life of Lazarus, as described in the eleventh chapter of John's Gospel.

But as recent Christian scholars note, such biological interpretations tend to subvert the real content and religious message these beliefs were intended to convey. Resurrection is not resuscitation, nor do any of the New Testament sources claim that Jesus' conscious, rational identity survived his death. Indeed, there is a consistent tradition that in the "meetings" between Jesus and his closest followers after the crucifixion, the "risen Lord" was a stranger whom they had great difficulty recognizing. In this sense, to say that Jesus was "raised from the dead" does not mean that he returned to haunt Jerusalem or Galilee as a terrifying, though familiar, ghost, but that he entered upon a new mode of existing, a new relation to God, a new and different way of interacting with the world.

Perhaps this revised interpretation of a traditional Christian doctrine opens the possibility of a convergence, Dr. Sperry, with what you have described about "transcendent mental content," the "very best of the conscious self" extended to exist beyond death. A related point of view was envisioned by Pierre Teilhard de Chardin, who emphasized the evolving spirituality of the human species:

It is done.
Once again the Fire has penetrated the earth.
Not with the sudden crash of thunderbolt,
riving the mountaintops:
does the Master break down doors to enter his own home?
Without earthquakes, or thunderclap:
the flame has lit up the whole world from within.
All things individually and collectively
are penetrated and flooded by it,

from the inmost core of the tiniest atom
to the mighty sweep of the most universal laws of being:
so naturally has it flooded every element, every energy,
every connecting link in the unity of our cosmos,
that one might suppose the cosmos to have burst
spontaneously into flame.

DR. SPERRY

Just as abandonment of the belief that the sun was driven
across the sky each day by the sun god Apollo subsequently
led to more sophisticated, more appealing theology, so also
with the called-for abandonment of dualistic concepts on
the one side, along with materialistic ideologies on the
other hand, one can hope and expect to see our belief
systems in the future evolve to higher, more sophisticated
levels.

The evolving spirituality of man has risen through progressive stages of increased insight and sophistication.

Recently changed views of mind-brain interaction carry
implications for all science. Traditional reductive materialist interpretations of science emphasizing causal control
from below upward are replaced by revised concepts that
emphasize the control exerted by higher emergent forces
from above downward. Conventional focus in science on
the role of material, mass-energy components in determining the nature of man and the universe is countered by an
increased emphasis on the crucial causal role played by the
non-material, space-time pattern or form factors.

The molecules and atoms of our world are seen to be
moved (their space-time trajectories determined) not so

much by atomic and molecular forces, as long predicated in science, nor by quantum mechanics, but rather by higher level forces such as manifest in biology, psychology, sociology, etc., that are not reducible in principle to the fundamental forces of physics. Mental and vital forces, long excluded and denounced by materialist philosophy, are reinstated in nonmystical form to their rightful role.

The whole concept of natural law as a foundation for moral judgment is significantly revised. Natural law can no longer be set apart from social, humanist or positivist frameworks, because it now includes these in the upper levels of a continuous hierarchic structure. On these new terms, a naturalistic or scientific theology is seen to yield a moral framework and outlook that has new credibility. This outlook satisfies spiritual and esthetic appeal, and at the same time promotes values that would appear to be of the type needed to counter current global trends toward worsening world conditions.

COMMENTARY

Sir John Eccles suggested at the beginning of this Conversation that in a world as perilous as ours is today, the decisive factor for the future will be human freedom, actualized as fully as possible and exercised responsibly. Does this view offer a solution to our worldwide crisis as you perceive it, Dr. Sperry?

DR. SPERRY

Partly. The one solution visible to date, in a way that would seem at all reasonable and humane, is to somehow achieve a change worldwide in the kinds of values and beliefs we live and govern by. This, of course, is where the need for a new theology or new global ethic comes in.

To halt or reverse the current population and other adverse trends is going to require counter forces of the most powerful kind. Nuclear war might do it, as might also a severe global famine, a large asteroid collision, or some other decimating worldwide catastrophe. The catastrophe from simply allowing present trends to continue should also be effective. A much happier solution is the one mentioned, namely a new value system, theology or global ethic that will bring a fundamental change in human value priorities. It would go a long way, for example, to help treat current world conditions if people generally were to acquire a deep conviction that it is not just unwise or inexpedient, but is actually immoral and even sacrilegious to pollute our world, to overpopulate, to deplete irreplacable resources, eradicate other species, or in any other way to despoil, degrade, or desecrate for coming generations the quality of our biosphere. Agreement that developments in this direction represent the logical, most promising key to a better future for our planet is now becoming widespread.

In my own case, the logic seems to carry to a deduction that the best way to get the needed new values and social priorities would be to achieve a union of religion and ethics with science. I should perhaps mention that the actual course of events and line of reasoning were the other way around, i.e., developments in science and value theory were seen to call for some revisions in the kinds of values and beliefs upheld by science. These in turn were perceived to be in a direction obviously suited to counter the adverse social trends.

It will be recognized that to propose a fusion of science with the value disciplines or to claim that developments in science support new social values is in both cases something that flies directly in the face of long established teaching regarding the relationship of science and values. The philosophic doctrine that it is logically impossible to derive values from scientific facts or to infer what logically *ought* to be from descriptions of what *is* has a venerable history extending back through G. E. Moore's *Principia Ethica* to at least Hume and some say to Plato. Attempts to find a basis for moral values in the natural order as described by science are customarily dismissed as examples of the "naturalistic fallacy."

In defense of our present position I contend that the traditional teaching that would keep facts separate from values and 'is' from 'ought' is itself based on a logical error. The error consists in assuming that values can be separated from brain function which by nature is intrinsically goal oriented and value guided. Human values, properties and products of brain function, cannot be treated with a pencil and paper logic that leaves the constraints of the functioning brain out of the picture. In brain processing, facts inevitably interact with and help to shape values.

For a simple shortcut to this revised fact-value logic consider the relation of values to belief. Most of us will readily agree that our values depend largely on the kinds of beliefs we hold, especially beliefs about the universe, about the nature of reality, of human consciousness, of the self, afterlife possibilities and so on. Most of us will also agree that science itself is a legitimate source of belief about these and other things and is, at the least, competitive with other sources such as intuition, revelation, authority, tradition, etc. In simple form the argument can be reduced to the reasoning that our values are shaped by science among other things. My proposal to fuse science and religion can be viewed largely as a proposal to meld and bring into harmony scientific and religious belief. It is not at all, of course, a proposal to start deriving or treating values directly by experimentation or by other scientific procedures.

COMMENTARY

Not just new ideas in science, but a new science and new scientists; not just new theories about human beings, but new human beings. These seem to be the core of Dr. Sperry's challenge. The future of human persons is, quite simply, the human persons of the future. These futuristic humans are not the strange polyurethane-coated robots of science fiction, but conscious living beings in whom higher emergent mental forces are freely working to discern and choose new social values. Neither science nor theology can be understood any longer as value-free enterprises in which the personal integrity of the investigator can be separated from the pursuit of "objective truth."

DR. JOSEPHSON

I'd like to go to something which is quite speculative but I think is indicated by the evidence. It is a correlation between the details of physics and the details of reality revealed by, e.g., the meditative experience. The correspondence is indicated as follows:

Correspondence Between Physical Reality and Subjective Experience

Physical Reality	Subjective Experience
Classical physics	World of sensory experience
Quantum physics	Celestial worlds
Unmanifest order	Transcendental experience

The ordinary reality as perceived by the senses corresponds to *classical physics*. The subtler realities of the astral or celestial worlds correspond to the aspect of physical reality described by *quantum physics*. Finally Bohm's unmanifest or implicate order corresponds to *transcendental experience*. These three experiential realities are experienced successively as one goes deeper into meditation. At first the mind concerns itself mainly with the ordinary aspects of life. Then deeper, non-ordinary experiences occur. Ultimately, these give way in turn to the silent, peaceful, free from any specific identifiable content (and hence, like the order postulated by Bohm, unmanifest), transcendental experience. Going back one step, the celestial worlds are only partly real; their status, according to mysticism, is that of fantasy or possibility, which reminds us of the possibility or potentiality aspects of the quantum wave function. I would like to suggest here an actual identity behind the parallelism.

COMMENTARY

Dr. Josephson, you seem to be arguing in a direction that proceeds in just the opposite manner from what Dr. Sperry has been suggesting. Sperry emphasizes proceeding "from the top downwards," since higher level forces enfold and supersede lower level laws such as those of quantum mechanics. But you seem to be arguing "from below upwards," from the subatomic level (quantum wave function) to higher levels where an unmanifest order exists. Is there a resolution for this opposition?

DR. JOSEPHSON

The picture one gets from sources like the Vedas is that our own intelligence is not something which is entirely within us: our inspirations and so on come from a different world, the celestial world. And this in turn is influenced by the absolute (or by God). So, to summarize, one can sketch out very roughly what the new kind of science, in which God is one of the actors, might look like. At the center we have this highest level of control, more or less what Dr. Sperry was talking about—the highest level of control would be conceived of as a very much more ideal and powerful version of our own intelligence. It would correspond to Bohm's unmanifest order, and would have profound influences on nature—in fact, the whole structure of quantum mechanics, on Bohm's model, is the result of the presence of this unmanifest order, and so the way we viewed science would be different. We would say really the whole structure of science once we get to the quantum level (where the unmanifest order makes observable order) would be directly the result of God's presence and works.

I must add this is my own interpretation of the theory; Bohm doesn't necessarily agree with this interpretation (or disagree with it). And we might hope that appropriate mathematical tools will be developed, so that in not too many years from now we'll have a new paradigm in which God and religion will be right in the middle of the picture, instead of being pushed out almost entirely as is the case at the present time.

Various sources such as the scriptures, or individuals who have evolved to an enlightened state, would provide information about how God goes about his works, whether he has other beings to help him, and what his motivations are and so on—some of this is in the scriptures and some is not. So, in other words, we can start to do what theologians are already trying to do, but treating it more scientifically, perhaps even developing appropriate mathematics. This mathematics might help theology by indicating limitations which a supreme intelligence might be subject to. This is very relevant because one of the difficulties religion has had is with paradoxes such as that of the existence of suffering in the world. And the kind of more scientific extension of theology might explain this by saying, for example, that God can do everything in principle but there are natural constraints which mean that certain things are impossible without disadvantageous side-effects, or, on the other hand, the constraints might mean that some processes would take a long time to be implemented. This kind of thing might be within the scope of the theological science of the future.

But the scientist is not keen on taking religious doctrine on authority. He likes to observe and to experiment. And here we get on to areas which are perhaps more controversial. We become involved not with religion as usually understood, but with mysticism. Mysticism is not widely appreciated even by people who regard themselves as religious, and so in order to avoid various kinds of misconceptions I should like to make first of all a number of preliminary comments.

I see mysticism and religion as things which are intimately related, and yet different from each other in important ways. I see mysticism as involving a more drastic change in the individual than is involved in ordinary religion. A mystic or someone trying to follow a path of enlightenment is trying (a bit like an athlete going through special training programs to enable him to be better physically) to follow a special training program (meditation or something similar) to enable him to have closer contact with the Absolute. And the mystic hopes to be able to see clearly into God's domain, whereas a person who doesn't go in for such intensive development may be in contact with God, but through a less direct channel. Now what one finds if one studies the various forms of mysticism is that the doctrines of the mystics are much less diverse than are religious doctrines. My interpretation of this is that mysticism is concerned with very fundamental laws—God and his relationship to man and other worlds beside the earthly world—whereas religion is a more applied sphere of activity: religion is concerned with the question, given that certain facts are the case, how should we live our lives? So the upshot of this is that I consider mysticism to be something universal like science, and that is the first point: and then the second point is that religions are based on the facts of this science. Thus mysticism is a kind of universal foundation for the diverse and different religions.

I should mention here that I'm not talking entirely about Eastern mysticism, because there is Western mysticism as well: e.g. Christian mysticism, Islamic mysticism (Sufism) and Jewish mysticism. These all say rather similar things.

After this detour, let me get back and say that a corollary of this is that mystical experience by self-development through meditation, etc., is not only the key to one's own development but also the key to understanding what is going on, the key to putting this attempt to synthesize science and religion on a solid foundation. And what I'd like to suggest is that if we follow this path of a synthesis of science with religion (using meditation as an observational tool), what we are doing is using our own nervous systems as instruments to observe the domains in which God works. Ordinary scientific instruments like telescopes, galvanometers and particle detectors are not going to be good in this context because they are designed to function in the material domain. Our nervous systems, on the other hand, are designed to allow us to interact not only with the material level of existence but also with the spiritual levels. And so all the different levels are open to exploration if we develop our nervous systems so that they tune in. One can imagine that this would be a part of the scientific training of the future.

❧   ☙

SUMMING UP CONVERSATION THREE

To repeat something Dr. Josephson just said, ". . . if we develop our nervous systems so that they tune in." Are we having a preview here of future humans? Are we saying that the new scientists studying human beings must develop themselves first as human instruments in addition to developing their instruments of steel and plastic and micro-electric chips?

This, our last Conversation, seems to have been dominated by a concern for the survival of the human species. In response to this concern, we have heard good news: we humans are at the tip of time's arrow, and we enfold all living things on this planet. Further, through will and intention we can change the results of so-called "immutable laws of the universe." Things are not running down, they are continuously becoming richer and fuller. We live in the first moments of a newly-becoming creation, not at its dwindling end. And we have the potential to make ourselves into the kind of future humans who will ensure a human future for all. Much disciplined work still needs to be done, of course, for the demands of a fuller, more palpably human integrity are neither easy nor easily met. We have yet much to discover about ourselves. But assistance in making such discoveries is becoming available. One thinks, for example, of Dr. Roy Laurens's recent *Fully Alive,* a book that grew experimentally from the sort of science promoted by these Nobel Laureates and that offers readers ways to discover and experience that fuller, more integral humanness hinted at in this final Isthmus Institute Conversation.

Though they would most certainly disclaim any evangelistic pretensions, Drs. Sperry, Eccles, Prigogine and Josephson have surely brought to public awareness a "gospel," a proclamation of good news. There is hope for a human future—and that future is nothing more or less than the humans who comprise it.

POSTLUDE

❧

A FINAL SUMMING UP

❧

BY NORMAN COUSINS

COMMENTARY

કેન્

NORMAN COUSINS

INTRODUCTION

In the following Commentary, Norman Cousins places these Nobel Conversations in a broader humanistic context. Mr. Cousins is almost unique in the world of American letters. He draws upon his experiences of over forty years —as long-time editor of the *Saturday Review;* as private citizen carrying out presidential missions abroad; as personal emissary for Pope John XXIII; as head of the project carrying out medical and surgical treatment for victims of the atomic bombing of Hiroshima; as head of a similar program that provided medical care for victims of Nazi medical experimentation; as organizer of a program that brought medicine and food to thousands of Biafran children; as one of the founders of public television in the United States; and as head of the special task force that set up an environmental program in New York City. He also serves on the medical advisory board of the Veterans' Administration hospitals.

After leaving the editorship of *Saturday Review* in 1978, he accepted an invitation to join the faculty of the School of Medicine at UCLA, where he now serves as adjunct professor in the Program of Medicine, Law, and Human Values.

In his own life and in his writings, Norman Cousins has demonstrated how mind and soul can be used as human tools to direct and realize the full potential of human beings.

MR. COUSINS

And how can we be certain, Lewis Carroll might have asked, that all that has been described or perceived by the philosophers and scientists may actually defy verification? Indeed, how do we know that the "world" itself may not be an illusion? Is it possible that all our vaunted scientific formulae are but speculations, and that we are but actors in a collective and unremembered dream?

The nature of reality, from the vantage point of different disciplines, is one of many rewarding exchanges in this volume. Proof of reality, of course, is dependent on laws of cause and effect. But the proof has to be perceived. Ultimately, therefore, everything is subjective. The radical idealists tell us that the forest itself does not exist except as it is experienced. We are forced to live with the nagging doubts that tell us that evidence of reality may be part of the illusion it seeks to negate. Yet we must work with the best that our senses have to offer. Even if reality is an assumption, it can be accepted as a working assumption and we are justified in assuming that causes have their effects and vice-versa. We have to be sustained by the conviction that the processes of inquiry, contemplation, and discourse are real and are as vital for the life of the mind as the products of farming are for the sustenance of the race.

In this sense, the exercise of reason can be even more impressive than its fruits. What is most challenging about the logical formulations of an Aristotle or a Bacon, or the orderly progressions of a Whitehead or a Russell, or the reflections on nothingness of a Sartre, is the fact of a human brain that can produce orderly progressions, reflect on nothing or everything, and zestfully explore new and better questions. A certain amount of philosophical pride can be derived from the use of the mechanism that can define and pursue its own aims. Indeed, the processes of thought are the ultimate wonder of the universe. Theories about creation, however beautifully fashioned or calibrated, are hardly more majestic than the thought processes that gave rise to them.

This book is a good example of the individual and collective manifestation of creative and systematic thought through the reaching-out of highly differentiated intellects to one another. Robert Maynard Hutchins spent more than half a lifetime fostering what he liked to call the "civilization of the dialogue." He would have found the exchanges in this book to be a substantial fulfillment of that aspiration. Hutchins, who regarded himself as a practical philosopher, was not given to ready exultations but he would have been deeply rewarded by the desire and ability of the participants in this exchange to hover over the junction where science, philosophy and religion meet in the modern world. For what they create is a totality that is more than the sum of its parts. It is not just the individual offerings of a Sperry or a Prigogine or an Eccles or a Josephson that go into a civilization of the dialogue; it is the combined interaction of their thoughts and the final edifice created thereby that are of primary value and significance.

This interaction helps to build high ground on which to experience a broader perspective and a wider vision. True, just at the point where great answers or solutions seem apparent, larger mysteries come into view. But this is no reason for deflection or despair. "If, after climbing one range of mountains," Norbert Weiner once said, "the physicist sees another on the horizon before him, it has not been put there deliberately to frustrate him." Nor is the fact that there is always another mountain range to be regarded as a damper on curiosity. When we widen our vision we widen our prospects. This process, John Dewey said, implies a unity with the universe: "A mind that has opened itself to experience and that has ripened through its discipline . . . knows that its wishes and acknowledgements are not final measures of the universe, whether in knowledge or conduct."

Obviously, what may be perceived as truth by probes in one field may be challenged by adjacent fields. But the eternal quest of science is to identify the vulnerabilities in any of its fields in order to clear the ground for newer approaches to truth. This quest is what helps us, in Vaihinger's phrase, to find our way around this world more easily. And this is the message that Einstein gave us in his analysis of Newtonian flaws and in his prediction that his own work would be superseded.

In effect, the exchanges in this work serve to relieve the loneliness and fragmentation inherent in serious investigation and thought. Scientific research, whether directed to the "properties of the living brain in action" or the "nonmaterial components of space and time," to quote from some passages in this book, doesn't always proceed out of a common center or lead to a common destination. Therefore, it is more than merely useful, it is imperative, to explore what Ruth Nanda Anshen in her new philosophical series calls "convergence."

Hans Zinsser, whose book *As I Remember Him,* has always seemed to me to be an example of convergence—of medicine, philosophy, and art—uses the word "coordinations" to express the same idea. He says that the greatest advances in science were not the result of bolts out of the blue or of explosive inspiration. Newton's experience with the falling apple, Descartes' geometrical discoveries while in bed, Darwin's flash of realization while reading Malthus, and Einstein's reflections of the Michelson puzzle while puttering around a patent office in Berne—all these contributions, Zinsser properly insists, were not just eruptions of genius, but "the final coordinations of innumerable accumulated facts and impressions which lesser men could grasp only in their correlated isolation but which by them—were seen in entirety and integrated into general principles."

This book is a superb effort in convergence and coordination. It deals with the furthest reaches of modern scientific theory but it does so by connecting new findings and theories to one another and by creating a pattern rather than a splattering of colors. In his history of science, George Sarton wrote about the importance of the interpenetration of ideas, not just between one scientist and another but between entire cultures and indeed civilizations. In this light, *Nobel Prize Conversations* is a specific and commendable service to the tradition out of which it came.

Roger Sperry, it seems to me, locates and reflects convergence and coordination when he says in these dialogues that "Among all the forces that impinge on mankind affecting our welfare and future, none is of more prominent and critical importance than the forces of human society by which we are surrounded and which, of course, are often personal, caring and replete throughout with purpose. The kinds of forces embodied in society, in family, in friends, politics, legislation, urban development and all the rest, including the expansion of ethical, moral, and religious values, are part of the natural order." To this he adds what may be the greatest of all subjective truths; namely, that evolution "can be viewed as a gradual emergence of increased purposefulness among the forces that move and govern living things."

In this proposition, I believe the other contributors to this volume all concur. What emerges from all their probes and ponderings is the endless challenge to improve the human condition. Science has provided the means for conquering squalor. But it has also provided the species with grisly devices to burn off all the accumulated works of creative splendor and, indeed, the human habitat itself. It is a commonplace to say we have the power to choose our destiny. But the prior question is whether we know how to choose or, more serious still, whether we are equipped to choose.

Here, the advice given by Sir John Eccles, if not a promise of that capability, at least points to an open door. He sees human freedom as the ultimate prize. His research has demonstrated to him that the same properties of the human brain that make thought possible can also comprehend the value of freedom and can devise ways of preserving life itself.

If this means that we have to do something that has never been done before—creating and sustaining a design for survival with freedom and justice—we need not despair. For what is manifestly unique about the human mind is that it can do things for the first time.

FULLY ALIVE

Readers of *Nobel Prize Conversations* may also find of interest a companion book entitled *Fully Alive,* by Dr. Roy Laurens. In this exciting and revolutionary book, Dr. Laurens tells his story both as a scientist and a very human man. The reader walks with him through his search for a larger experience, one that leads to an expansion of reality far beyond his dreams. *Fully Alive* is at once a wonderful personal story, an important contribution to the growing literature on the nature of the mind, and a clear and simple self-help guide to awakening potentials for achievement in business, health and personal relationships.

Members of the medical profession have enthusiastically endorsed this book. Dr. James Hall, who contributes the book's "Introduction," calls it "possibly a fundamental revolution, a profound theoretical insight into personal scientific experimental technique."

NOBEL PRIZE CONVERSATIONS

୬

NOTES

NOTES ON THE TEXT

The following notes provide further information about the poetry, prose and other literature quoted by participants in The Isthmus Conversations.

PRELUDE

The reflections prompted by this chapter's opening "Hymn," by A. R. Ammons, lead to an imaginative reconstruction of the atmosphere surrounding both the Nobel Prize ceremonies in Stockholm and the Isthmus Institute lectures in Dallas. On the common ground of fantasy, powered by the human imagination, poets meet scientists and novelists meet nuns. Nelly Sachs, winner of the Nobel Prize for Literature in 1966 is represented with a quotation from her poem "O the Chimneys." She shares imaginary space with Ernest Hemingway, another Nobel Laureate in Literature (1954), and with Albert Einstein, Laureate in Physics (1921). Sachs's lamentations pour out of the unutterable horror of war and holocaust, and are echoed in the life of

Mother Teresa, whose work with the starving poor of India was acknowledged by the Prize for Peace in 1979.

Represented too in this "Prelude" are the four Nobel Laureates whose conversations at the Isthmus Institute are presented in this book. Readers who may want to acquaint themselves better with these men and their ideas about human mind and human future are encouraged to read some of the following:

Roger Sperry's *Science and Moral Priority* (New York: Columbia University Press, 1983) persuasively argues the case for a union between science and values, on the premise that scientific experiment is never a value-free enterprise, nor scientists a tribe of ethically neutral observers who merely record and measure nature. Sperry, who won the Nobel Prize in Medicine/Physiology in 1981, believes that our future requires a fundamental change in human value priorities rather than a revival of our tendency to treat global ills by applying more science and technology. Some of the scientific reasons why such a change in values is required are documented by Karl Popper and Sir John Eccles in their book *The Self and Its Brain* (New York: Springer International, 1981). Further implications of Eccles's research into the brain's "supplementary motor area," that small "point" at the very top of one's head where electro-chemical and neural events are initiated by non-material mental intentions, are drawn out in his *The Wonder of Being Human* (New York: The Free Press/Macmillan, 1983).

Ilya Prigogine's vision of a universe of "becoming" rather than "being" is powerfully sketched in his book *From Being to Becoming* (San Francisco: W. H. Freeman and

Company, 1980). Winner of the 1977 Prize for Chemistry, Prigogine is convinced that *time* is our key to a new dialogue with nature. "We are in a period," he writes, "in which we have to transcend the classical opposition between what is *in* time and what is *out* of time." Our modern search, Prigogine concludes, is "for a junction between stillness and motion, time arrested and time passing." These ideas are developed still further in Prigogine's *Order Out of Chaos* (New York: Bantam Books, 1984), written in collaboration with Isabelle Stengers.

Ilya Prigogine's invitation to transcend the classical opposition between what is in and out of time gives wings to the opening "Prelude." Presences from past Nobel ceremonies crowd the imagination. Physicists Max Planck (1918) and Niels Bohr (1922) are there, as are novelists Sigrid Undset (1928) and Saul Bellow (1976). Undset's trilogy *Kristin Lavransdatter,* written between 1920 and 1922, remains popular in the English translation by Charles Archer and J. S. Scott (*The Bridal Wreath; The Mistress of Husaby; The Cross;* New York: Alfred A. Knopf, 1923–1927). Saul Bellow, of course, continues to enrich the world's literature with recent novels such as *The Dean's December* (New York: Harper and Row, 1982). In the United States, the breakthroughs in theoretical physics achieved by Nobel Laureates like Max Planck, Niels Bohr and Albert Einstein have been made accessible to many nonspecialist readers by Isaac Asimov's popular books, especially his recent *Asimov's New Guide to Science* (revised edition; New York: Basic Books, 1984).

Though he never won the Nobel Prize for Literature, American poet Wallace Stevens is another presence encountered in the opening "Prelude." Lines from his poem

"The Snow Man" are quoted in the description of Roger Sperry. Stevens's older contemporary, Irish poet William Butler Yeats, who won the Prize in 1923, is cited in the same context, where lines from "Sailing to Byzantium" appear. Still another literary figure, a fellow-countryman of Dr. Brian Josephson, appears briefly later on in the "Prelude," where allusion is made to J.R.R. Tolkien's "middle earth." Though, like Stevens, Tolkien never received a Nobel Prize, his epic trilogy *Lord of the Rings* continues to excite the imagination of new generations of readers, especially in its popular paperbound edition (New York: Ballantine Books, 1965).

Finally readers should note that the excerpts from the Nobel Prize presentation lectures by Professors David Ottoson, R. Granit, Stig Claesson and Stig Lundquist—all of the Karolinska Institutet in Stockholm—are printed here by kind permission of the Nobel Foundation, which has also granted permission to reproduce the photographs of Drs. Sperry, Eccles, Josephson and Prigogine.

CONVERSATION ONE

The power of the never-resting mind, as it acts through mental intentions, thinking and decisions that enfold and supersede the purely material functions of the brain, forms the central theme of Conversation One. The theme is sustained principally by the Isthmus Institute exchanges between Roger Sperry and John Eccles, but it is further elaborated by the commentary woven throughout the conversation. The result is a dialogue between scientists and humanists who unite to stress the priority of human values within the field of scientific experimentation generally and mind/brain research specifically.

Unexpected partners join the conversation between Sperry and Eccles from time to time. Emily Dickinson is heard from in a quote from her poem "I felt a Funeral in my Brain." The quotation intends no pun, though it is true that both Sperry and Eccles celebrate a requiem for the outmoded materialist view which excludes mental realities from the domain of "pure science" and discounts the existence of mind distinct from brain. Allen Ginsberg's "Howl" is quoted in the same breath with Dickinson. Both these poets, though separated by a century and by enormous differences in diction and technique, reach a level of intuitive "rightness" in their work which permits us to catch the never-resting mind as it reaches beyond the limits of the brain's material functions. Dickinson's *Complete Poems* are, happily, available now in the critically acclaimed edition by Thomas H. Johnson (Cambridge, Mass.: The Belknap Press of Harvard University, 1983), while Ginsberg's *Howl and Other Poems,* introduced by the late William Carlos Williams, is still available from City Lights Books, San Francisco (1959; 30th printing, 1980), or in his *Collected Poems 1947–1980* (New York: Harper and Row, 1984).

A bit later in this conversation, Wallace Stevens reappears. In a passage from his "Sunday Morning," Stevens contends that the human mind is potent enough to enfold time passing. Thus, even when earth's satisfactions disappear, "April's green" really *endures* in the human memory, just as the "desire for June and evening" truly *lives* by the mind's power of anticipation. As Sperry might put it, memory and anticipation are real mental acts that enfold time and supersede—though they don't replace—the brain's electrochemical activity.

Joining the conversation later on is America's first poet of worldwide stature, Walt Whitman. Section 33 of his "Song of Myself" is quoted to illustrate the imagistic connections that arise in the brain's right hemisphere. As Sperry and Eccles note, the right brain does not think in linear, logical sentences, but works rather in whole images and single, symbol-laden words. It matters little whether these images "make sense" on the level of rational analysis or discursive reasoning. Whitman's almost surreal catalogues of daily activities and natural wonders seem to anticipate modern research about the right hemisphere's mode of grasping reality in whole patterns. Whitman's work is cited from the handsome new edition of his *Poetry and Prose* published in New York by The Library of America (1982, pages 220–221; 554–556).

As the first conversation draws to a close, more lines from Wallace Stevens's "The Poems of Our Climate" are quoted, as we listen to the poet struggle with the rational mind's compulsion to control reality through its disorderly passion for words. Finally, in the "Summing Up," Dr. Carl Sagan's interview with editors of *Parade Magazine* (December 2, 1984) is quoted in confirmation of the intimate kinship that unites all living things in our evolving universe.

CONVERSATION TWO

The second Isthmus Conversation highlights an exchange between Roger Sperry and Brian Josephson. As

Sperry and Josephson call our attention to the relation between mind/mental activity and the larger, evolving universe of matter, time and space, Walt Whitman is heard from once again. Section 32 of his "Song of Myself" is quoted to illustrate Sperry's conviction that vital forces, living powers, inhabit the species that populate our planet, and that these forces cannot simply be reduced to the laws of quantum physics.

Roger Sperry's argument that strictly materialist science cannot fully account for the life and activity we see and touch in our universe is supported as well by Brian Josephson, who refers to a recent work by Richard L. Thompson entitled *Mechanistic and Non-mechanistic Science* (Lynwood, New York: Bala Books, 1981). Thompson discusses in some detail the difficulties conventional science encounters when it attempts to explain phenomena like evolution and creativity. Dr. Brian Josephson also refers to works by Fritjof Capra and David Bohm. In *The Tao of Physics,* Capra explores the connections that may possibly link modern scientific discoveries with features of mysticism (Berkeley, California: Shambhala Publications, 1975). David Bohm's *Wholeness and the Implicate Order* (London and Boston: Routledge and Kegan Paul, 1980) discusses some of the paradoxes of quantum mechanics and stresses the existence of "hidden variables" which affect physics but cannot be observed directly. These variables, Bohm contends, reveal the existence of an "implicate order" in the universe which cannot be directly measured, even though we can see its results in the observable phenomena that surround us.

INTERLUDE I

The short poem which introduces "Interlude I," an edited collection of comments from guests at the Isthmus Institute, is taken from Rainer Maria Rilke's "A Book for the Hours of Prayer," in a new English translation by poet Robert Bly (*Selected Poems of Rainer Maria Rilke,* translated, with commentary, by Robert Bly. New York: Harper and Row, 1981, page 13).

INTERLUDE II

Following the editorial "Introduction" to this second "Interlude," a quotation from Karl Popper's book *The Logic of Scientific Discovery* (London: Hutchinson, 1977; preface to the 1959 edition) opens Dr. Prigogine's remarks. Early on, Ivor Leclerc's *The Nature of Physical Existence* (New York: Humanities Press, 1972) is cited to support Prigogine's argument that most modern European philosophy has had to deal with the tragic choice between the mechanistic view of the world supported by classical physics and our own daily experience of the irreversible, creative dimension of life. This central philosophical problem received perhaps its most influential formulation in Martin Heidegger's *Being and Time* (translated by John Macquarrie and Edward Robinson; New York: Harper and Row, 1962).

Readers who wish to explore some of the more technical aspects of Prigogine's discussion of time, especially as it relates to the work of earlier twentieth-century scientists such as Max Planck and Ludwig Botlzmann, may consult the following works: Fritz Rohrlich, "Facing Quantum Mechanical Reality," in *Science* 221 (1983), 1251–1253; Ilya Prigogine, *From Being to Becoming.* Time and Complexity in the Physical Sciences (San Francisco: Wm. H. Freeman, 1980); B. Misra and I. Prigogine, "Time, Probability and Dynamics," in *Long-Time Prediction in Dynamics,* edited by C. W. Horton, L. E. Reichel and A. G. Szebehely (New York: Wiley, 1983).

The riddle of time, as it troubled a great physicist like Albert Einstein, is noted once more in Prigogine's essay with a quotation from *The Philosophy of Rudolf Carnap,* edited by P. A. Schlipp (Cambridge: Cambridge University Press, 1963).

Toward the conclusion of his study, Prigogine turns to the poets Paul Valéry and T. S. Eliot. Valéry's notebooks are quoted (*Cahiers;* Paris: Gallimard, I, 1303), as are Eliot's *Four Quartets* ("Burnt Norton," I; *The Complete Poems and Play, 1909–1950;* New York: Harcourt, Brace, Jovanovich, 1971, page 117). Both these poets seemed to anticipate modern science's conclusion that time cannot simply be reduced to quantitative, airtight compartments labelled "past," "present," and "future."

Finally, Prigogine calls our attention to Alfred North Whitehead's *Process and Reality* (Cambridge: Cambridge University Press, 1929) and to George Steiner's study of *Martin Heidegger* (New York: The Viking Press, 1978). It was Whitehead, perhaps more than any other modern philosopher, who insisted that no being—including God—can be defined apart from activity, that merely passive matter cannot lead to a creative universe.

CONVERSATION THREE

The third and final Isthmus Conversation addresses an issue important to all: the human future and future humans. The voice of Emily Dickinson, heard earlier in these conversations, echoes again. Dickinson's astonishing "He fumbles at your Soul" subverts common assumptions about religious experience and invites reflection on new ways to perceive the relation between humans, "God" and cosmos (*The Complete Poems of Emily Dickinson,* edited by Thomas H. Johnson; Boston: Little, Brown and Company, 1960, page 148).

A quotation from Jesuit priest and palaeontologist Pierre Teilhard de Chardin appears toward the end of the third conversation. A scientist who was involved in the discovery of "Peking man," Teilhard de Chardin became convinced that humans are the axis of cosmic evolution and that the future of the universe is future humans. Evolution, he felt, will reach its final stage when all individual human consciousness is integrated in the "Omega Point." These, and related ideas, are developed in his posthumously published works *The Phenomenon of Man* (introduced by Julian Huxley and translated by Bernard Wall; New York: Harper, 1959) and *The Hymn of the Universe* New York: Harper and Row, 1965). The quotation cited here is taken from *The Hymn of the Universe.*

ACKNOWLEDGMENTS

All the Nobel Prize Conversations, as well as the Prelude, Interludes and the Postlude in this book, are introduced by quotations from poetry or prose. Similarly, the edited introductions and commentaries include a number of lines or stanzas of poetry. Grateful acknowledgement is made for permission to quote:

PRELUDE

The lines from "Hymn" from *Collected Poems, 1951–1971*, by A. R. Ammons, by permission of W. W. Norton and Company, Inc. Copyright © 1972 by A. R. Ammons.

The lines from "O the Chimneys," translated by Michael Roloff, from *0 The Chimneys*, by Nelly Sachs, translated by Michael Hamburger, Christopher Home, Ruth and Matthew Mead, Michael Roloff; by permission of Farrar, Straus, and Giroux, Inc. Copyright © 1967 by Farrar, Straus, and Giroux.

The photographs of the Nobel Laureates and the portions of speeches made at the Nobel Prize ceremonies by Professors

CONVERSATION ONE